Navigating the Cosmos: A New Framework for Assessing Compliance with International Space Law

Nimra

Copyright © 2024 by Nimra

CONTENTS

1 INTRODUCTION

As governments, citizens, and corporations increasingly engage in outer space activities, the governance of behaviour according to internationally accepted laws and norms remains paramount now more than ever to ensuring the continued access to and use of space for all. While some countries may subscribe wholly to these laws and norms, others may ignore them or actively attempt to subvert them for various economic or political reasons. These conditions therefore create the need for a strong, public understanding of all spacefaring states' behaviour for monitoring their adherence to the very same international laws and norms that made their own access to and use of space possible. This book proposes a contribution to the field of space policy and governance in an effort to address a gap in the evaluation of states' behaviours in outer space. A comprehensive, periodic, and systematic measure of states' efforts to comply with international space law and normative behaviour does not exist, suggesting a critical need to ensure robust and informed debate around countries' activities in outer space. Therefore, this book endeavours to propose a set of criteria comprising a framework for which space policy stakeholders can apply to countries and develop a comparative understanding of their levels of compliance with international space law and normative practices with regard to outer space. This contribution conceptually sits among the increasing body of work on ratings and rankings of countries on a variety of issue areas ranging from health and environment to democracy and human rights, in addition to the evidence-based policy and programming movement.

As democratic governments and states developed with normative governance processes as part of their political consolidation and development throughout history, so did too elements of society in the form of civil society organisations that strove to hold them accountable and responsive to the needs and interest of the population at large. Contemporary understandings of civil society as antagonists to state governments and promoters of political change came about during the social movements against Eastern European communist states during the 1970s and 1980s; however, Gordon White offers a more inclusive and conceptually useful approach for understanding what is meant by civil society in the present day: "an intermediate associational realm between state and family populated by organisations which are separate from the state, enjoy autonomy in relation to the state and are formed voluntarily by members of society to protect or extended their interests or values," (Burnell and Calvert 2004). This book therefore employs this approach for framing and understanding civil society as a potential user group for the framework proposed in later chapters.

In particular, given the advent of international governance mechanisms and regimes, such as the United Nations, the growth of civil society as a critical and accountability-seeking voice developed in this sphere too. This area witnessed the growth of international and cross-national civil society groups seeking to not only advocate at the international level toward the corresponding international institutions, but also as a means of encouraging domestication of international law and norms among domestic institutions. The Universal Periodic Review (UPR) Process, for example, was established in 2006 through which countries review each other's human rights record in an effort to improve the human rights environment in each. Civil society organisations play an important role in the review process, and help to provide informed and critical analysis of each state's record by participating in the review meetings, providing background information, advocating in bilateral and multilateral meetings with the state reviewers, and monitoring implementation of UPR recommendations. Moreover, the existence of a platform such as the UPR has allowed civil organisations the opportunity to leverage the international happenings at the UPR (i.e. state-to-state peer review) for the benefit of domestic advocacy initiatives. As an example of this, Eric Tars and Deodonne Bhattarai (2011) cite the case of housing in the United States following that country's 2010 review. The UPR review led to a nationwide consultation including a year-long assessment of human rights by nongovernmental organisations (NGOs) that resulted in reports of findings supplied to the Obama Administration and U.S. government, which in turn on March 18, 2011 reacted positively in advancing the rights to housing for citizens.

Therefore, with the growth of international governance regimes and normative practices for states to follow, the governance and use of outer space sits as an emerging area in which similar critical and accountability-seeking dynamics must evolve given its relatively nascent stage of development, as compared to other areas in which international governance mechanisms exist such as human rights, nuclear non-proliferation, and trade, among others.

Similarly, as seen with the growth of national and international civil society groups, the use and implementation of evidence-based advocacy techniques accompanied their rise to improve their policy advocacy effectiveness and legitimacy among policymakers. Included as part of their use of evidenced-based advocacy, civil society has strived to integrate data-based and quasi-academically rigorous and sound methodological and evaluation techniques to provide for this evidenced-based advocacy. Numerous civil society groups, such as Transparency International or Freedom House, largely including institutions based in the Global North advocating for democracy, human rights, and good governance, prove as some examples whereby groups developed standard methodologies and indicators and established ways of measuring against such indicators on a periodic basis so as to develop comparable, historical datasets tracking progress in the areas in which they monitor.

This research would prove remiss to suggest such types of reviews and methods are not applied to assess activities in outer space, however, the example of the *Space Security Index* (2018) only measures such activities within a limited scope and understanding of the space environment. This assessment of space security, in addition to a review of contemporary ratings and rankings system implemented by civil society organisations, is discussed in subsequent sections of this book.

1.1 Objectives of this research

The research undertaken here is meant to explore, develop, and describe a framework whereby states' behaviour in outer space is assessed according to the ratified outer space treaties in addition to established consensus-based norms developed by the UN Committee on the Peaceful Use of Outer Space that have evolved through the ongoing use of outer space by the respective states. Moreover, this framework will be developed with a purposeful Global South perspective, emphasising equitability and inclusivity of perspectives and interests as to which elements of the framework should be prioritised in the assessment. The framework will strive to incorporate views from governmental and intergovernmental bodies, industry, academia, and civil society.

As described above, contemporary tools and methodologies exist for measuring and tracking compliance with internationally accepted standards of practice established in formal instruments of international law such as treaties or established in practice as norms.

Many of the existing tools and methodologies for monitoring governmental behaviour and compliance with standards and norms exist in an attempt to promote and create awareness of states' compliance with their obligations under international law. Several of these tools are found in the democracy, human rights, and governance (DRG) field in an effort to generate action and advocacy toward increased adherence and compliance with internationally accepted standards and norms. Employing a Theory of Change methodology, as an example of one such tool, in their programming logic, many of the organisations operating in the DRG sector rely on these tools and methodologies, and their application therein, to identify, in an evidenced-based approach, gaps and deficiencies in governmental behaviour that would then be used by citizens and organised groups alike to advocate for improvements in those areas that were assessed. The framework proposed in the following chapters follows a similar approach.

1.2 Evidence-based programming and policymaking

As noted previously, this book emerges from two historical movements, namely the evidence-based policy and programming movement and the accompanying rise of international ratings and rankings research. The following sections explore these trends and identify, where applicable, their relationships to the proposed framework assessing countries' levels of compliance with international space law and the challenges and potential pitfalls such a framework should avoid in its development and application.

Evidence-based policy, and thus evidenced-based programming, emerged in the 1980s and 1990s. Cornish (2015) notes that "Emerging in the mid-1980s, the Evidence-Based Policy (EBP) movement has gained supporters, become a commonplace in many disciplines, and become a gatekeeper for the allocation of intervention funding." Cornish (2015) continues "The important claims [for evidence-based policy] were claims about a science that is empirical (i.e. based on examination of the world rather than hearsay or ideology), and which is disinterested (i.e. independent of political influence)." Foss Hansen (2014) builds on this by writing of the developing evidence-based policy movement emerging from the field of medicine to other fields such as public health, social care, education, and international development. As suggested here, the evidence-based movement emerged from the health and medical disciplines during the 1980s and early 1990s and shifted to in the social science and development fields as well. Evidence-based programming used by international development organisations, including those across the spectrum ranging from health and agriculture to democracy and human rights, has increased over the past decade. Craig Valters (2014) argues that social practice has striven to be informed by evidence and that practitioners have sought out and implemented tools based on prevailing development theories.

In particular, this has been due to the push for increased monitoring and evaluation demands imposed upon these groups, especially as resources, often publicly funded, require justification for their expense in the political arena. Especially regarding United States foreign assistance, Ferrarello (2017) highlights "There's a misconception about U.S. spending on foreign assistance. While public opinion polls suggest that Americans believe foreign aid makes up roughly 25 percent of federal spending, it actually makes up a mere one percent of the federal budget." Politicians in the United States, and elsewhere, have often targeted foreign assistance due to these prevailing misconceptions among the public, including President Donald Trump who, Ferrarello (2017) notes, sought to reduce foreign by 37 percent and merge the United States's principal international development agency, USAID, with the Department of State, a more diplomatic and politically-oriented organisation within the U.S. government. This is supported when Nermine Wally (2018) of the Cairo Institute for Liberal Arts and Sciences notes: "Both beneficiaries and donors of international aid have increasingly been pressured by their partners and constituents to capture what works, how it works and for whom it works in a both rigorous and systematic way. Donors and governments alike are expected to utilise the findings emerging from evaluation in their decision-making process to justify resource mobilisation." Harry Jones of the Overseas Development Institute (2012) supports this viewpoint as he writes of high stakes when it comes the need to improve learning efficiently and cost-effectively while also making the results of learning and evaluation relevant and useful to actual practice.

Therefore, organisations, including those in the democracy and human rights field, have moved toward evidence-based programming in order to perform their work more effectively and achieve results, while simultaneously responding to donor demands for greater justification of their work. According to a London School of Economics report, "Theories of Change are increasingly mandatory for implementing agencies to submit to donors in the aid industry," (Craig 2014).

For example, the United Kingdom's Department for International Development (DFID), has been using the theory of change methodology in its work since 2010 in order to better inform DFID development approaches and efforts. In a 2012 review of DFID, the importance of theory of change and evidence-based programming in the department's work was highlighted as follows: "Theory of change thinking is used to explain rationales and how things are intended to work, but also to explore new possibilities through critical thinking… Critical thinking is cross-checked with evidence from research (qualitative and quantitative) and wider learning that brings other analytical perspectives, referenced to stakeholders', partners' and beneficiaries' contextual knowledge" (Vogel 2012). Similarly, the United Nations Development Group (2018) views its approach toward theory of change as a programming methodology in the following way: "A theory of change can help a United Nations Country Team (UNCT) systematically think through the many underlying and root causes of development challenges, and how they influence each other, when determining what an UNDAF should address as a priority to maximize the UN's contribution to achieving development change."

As seen here with the further introduction of a requirement for rigorous tools and methodologies to develop and guide programming, organisations have responded through the use of surveys and other forms of social science research, largely for those operating in the democracy and human rights field. In an attempt to make their programming more sound and legitimate, surveys and ratings and rankings have gained influence. Although some of these approaches and surveys, such as the annual Freedom House *Freedom in the World* report, have been in existence for decades, others are relatively new, such as the *CIVICUS Monitor*. The general use of ratings and rankings by these organisations is explored in the upcoming section, followed by an overview of the most popular and well-known of these international rankings.

1.3 Ratings and rankings research

In addition to this book being situated among the evidence-based movement for programming, it also sits well within efforts to provide insights and research on international and country-level processes through ratings and rankings research. Several such organisations and surveys of these types exist presently, although some have existed longer than others. Academics and policymakers alike have argued for and against these international ratings and rankings, demonstrating that advantages and disadvantages exist for each. However, generally, these approaches, also called scorecards, are used to measure how countries or other actors perform in certain policy areas.

Because of the ease with which these reports can be consumed by policymakers and the public, they are often used to pressure countries into being ranked; hence, countries often take these reports seriously. According to Judith Kelley (2017), "For example, if your country is at the bottom of a well-respected scorecard for "Ease of Doing Business," you might find that international businesses start to avoid investing in your economy." Sarah Sunn Bush

(2017) agrees with Judith Kelley when she writes that scorecards are influential toward improving countries' behaviours. As such, more than 180 ratings have been developed to measure country-level performance in different issue areas. Robert Gregory (2014) supports this observation in the growth of the measures used to evaluate countries: "Apart from the WGI [World Governance Index], there has been an explosion of indexes and indicators, as various international organisations develop measures to rank comparatively the performances of different countries, both globally and regionally." These measures are used as a form of social pressure by countries and organisations alike to influence each other's behaviour as the exercise of military or economic sanctions diminishes.

Compared to the advent of evidence-based programming described previously in this book, which emerged in the 1980s and early 1990s, the majority of measures of countries' performance developed after 1990 despite their existence, such as in the form of sovereign credit ratings, since the 1930s. Judith Kelly and Beth Simmons (2014) argue that this explosion in ratings and rankings has been due to various contemporary trends:

"Meanwhile, the cost of exerting pressure via information has declined. While not costless, it has never been easier to collect and distribute reasonably credible information from highly decentralized sources on a global scale than it is today. Moreover, the indicization of information is a natural response to demands for transparency and accountability (Mathiason 2004). It is likely that the convergence of normative prohibitions against overt force and the ease of collecting, analysing, and disseminating information globally has encouraged the turn to indicators as tools of international influence."

Therefore, ratings and rankings serve as a more cost-effective and internationally-acceptable way of exerting pressure outside of military and economic action, thus constituting a form of soft power countries are able to exercise. The move toward using ratings and rankings to exert pressure has also been driven in part by "a combination of rational interests, market demand, and institutional design. This is most notable in relation to private market governance, where one of the main motivations behind benchmarking has been to produce useable information that improves how actors respond to market forces and conditions," (Broome and Quirk 2015). The rating system to be proposed later in this book falls squarely in this growth and accepted use of these measures of countries' performance: "Benchmarking efforts now play a key role in policy coordination and institutional design among states and IOs [international organisations] faced with collective action problems over climate change, disaster, management, and human development," (Broome and Quirk 2015).

With the growth of ratings and rankings, however, there have been various areas of concern and critique toward some of these instruments in use. Gregory Michener (2015), in an examination of international transparency policy indexes, specifically highlights some of these challenges when he writes:

"The driving question here is whether the design of policy evaluations, particularly those using index-based formats, is motivating "the right type of compliance". Impressive scores across several international transparency policy indexes (ITPIs) by countries with uneven, if not questionable institutional track records understandably raise questions about how easily "gamed" these measures might be (Eisenkopf, 2009; Hood, 2012, pp. S86–S88). In other words, what (avoidable) "loopholes" might indexes afford by reason of flawed evaluative strategies, designs, or challenges associated with the evaluation of policies via the composite index format?"

Nelson Espeland and Michael Sauder (2007) point to a further complication with ratings and rankings as they note that those being measured alter and modify their behaviour in response to being measured and observed through the process of reactivity and reflexivity. Indeed, the intended purposes of many ratings and rankings, particularly those used by international advocacy organisations, is in fact to modify behaviour of those being measured and to equip advocates and other policymakers with the evidence they need to pressure for change. Despite the tension here, Espeland and Sauder (2007) do highlight an important phenomenon that occurs when those being measured know they are being measured, namely playing to the test and trying to improve scores on indicators themselves without actual improvements in on-the-ground conditions. Therefore, although advocates and consumers of rankings information may encourage those countries being measured to undertake efforts to improve their behaviour (assuming it is deficient or non-compliant in some regard), they may encounter the risk of countries merely following a rote exercise of conforming to the measurement parameters without undergoing deep or substantive reform efforts.

1.4 Structure of this book

Given the extent to which ratings and rankings systems, and the organisations that implement them, have come to characterise much of the political economic understanding of the world, it remains important in Chapter 2 to explore some of the most well-known of these and the methodologies behind their assessments. Thereafter, this book continues with a discussion of a select group of these examples to illustrate challenges associated with these, some of which have been referenced in the preceding section and which will prove important for consideration in the development and description of the framework to be proposed for assessing compliance with international law and norms regarding the use of outer space. Chapter 3 then moves on to explore the existing set of international laws and norms relating to outer space, and discusses which of these will be used for informing the proposed framework. Chapter 4 will seek to understand the various elements comprising assessment methodologies and which to use based upon identified best practices contained in previous research. Chapter 4 helps ground Chapter 5, wherein the proposed framework is described in detail accompanied by questions for each indicator and a scoring guide to assist with arriving at a comprehensive assessment score. Chapter 6 concludes the book by exploring the potential scores resulting from the application of the framework to the three test cases (the United States, South Korea, and South Africa), and the various ranges and categories (i.e. highly compliant, mostly compliant, partially compliant, minimally compliant, or non-compliant) in which these scores fall for purposes of advocacy and decision-making.

2 EXISTING RATINGS AND RANKINGS SYSTEMS

One such example of a ranking and ratings system sits with the global civil society alliance organisation CIVICUS in the form its *CIVICUS Monitor*. This system rates countries according to five categories, including closed, repressed, obstructed, narrowed, and open, which are used to assess civil society freedoms across the world. As CIVICUS describes in its methodology report on the system, it relies on civil society-produced reports on civic space, international indices compiled by international civil society organisations, CIVICUS-produced and country-specific reports, narrative and quantitative reports produced by the group's in-country partners, and user feedback. Based upon a system of weighting of each of these inputs and review by independent experts, a score is produced on a range of 1 to 100 with each category classification corresponding to a numerical score (for example a 'closed' country sits in the score range of 1 to 20).

Freedom House, a United States-based organisation, produces several ratings and rankings reports; however, the most well-known of these is the annual *Freedom in the World* report, which classifies countries as free, partly free, and not free according to scoring of political rights and civil liberties. A country's score, which translates into to its category classification, is determined by an expert analysis who writes a narrative report and scores the country based upon a series of questions, assigning each question a numerical value from 0 to 4. There are two main groups of questions used by Freedom House to arrive at its overall scores for countries, including political rights and civil liberties questions. Each of these are then further divided in to subgroups, including electoral process, political pluralism and participation, and functioning of questions for the political rights group and freedom of expression and belief, associational and organizational rights, rule of law, and personal autonomy and individual rights questions for the civil liberties group. The questions in each of these groups and subgroups are based on the Universal Declaration of Human Rights adopted by the United Nations in 1948, demonstrating that the assessment is a normative assessment of countries based on existing and accepted international law and understandings of rights. However, Freedom House (2018) clarifies in its methodological description that it does "assess the real-world rights and freedoms enjoyed by individuals, rather than governments or government performance per se. Political rights and civil liberties can be affected by both state and nonstate actors, including insurgents and other armed groups."

Transparency International, an international civil society organisation promoting good governance, developed another well-known ratings and rankings system, the *Corruptions Perception Index* (CPI). The CPI is a composite indicator that was developed in 1995, and which was revised in 2012 in order to allow historical comparison of countries' scores over time. The CPI develops its corruption indicator based on 13 sources of data that address questions of bribery, diversion of public monies, use of public office for private gain, civil service nepotism, and state capture by elites. A country is assigned a score based on an average of at least three of the 13 data sources, which is made possible by recalibrating the data sources own scores on to a 0 to 100 range. This rescaling of the source data is necessary as the various sources do not follow the same ratings and rankings approach when developing their own individual indicator scores. The CPI 2017 relied on these 13 data sources and their indicators, some of which are discussed in more detail in this book, including: the African Development Bank Country Policy and Institutional Assessment, Bertelsmann Stiftung Sustainable Governance Indicators, Bertelsmann Stiftung Transformation Index, Economist Intelligence Unit Country Risk Service, Freedom House Nations in Transit, Global Insight

Country Risk Ratings, IMD World Competitiveness Centre World Competitiveness Yearbook Executive Opinion Survey, Political and Economic Risk Consultancy Asian Intelligence, The PRS Group International Country Risk Guide, World Bank Country Policy and Institutional Assessment, World Economic Forum Executive Opinion Survey, World Justice Project Rule of Law Index Expert Survey, and Varieties of Democracy.

The World Bank, in collaboration with the United States-based think tank Brookings Institution, produces another commonly used ratings and rankings system, the *Worldwide Governance Indicators* (WGI). As the name suggests, the WGI have addressed six areas of governance for countries and territories since 1996. The areas of measurement include voice and accountability, political stability and absence of violence, government effectiveness, regulatory quality, rule of law, and control of corruption. Similar to Transparency International's CPI, the WGI are a composite of indicators based on over 30 existing data sources, which include four different types of data such as surveys of households and firms, commercial business information providers, non-governmental organisations, and public sector organisations. In fact, the WGI indicators rely upon data provided by the Heritage Foundation, Freedom House and Transparency International, in addition to base data sources that already serve as inputs to Transparency International's own assessment (for example, the Varieties of Democracy Project and the Economist's Intelligence Unit). This suggests that some sources of data have outsized influence, and therefore suggests problematic methodological issues, within the ratings and rankings ecosystem. While an exploration of these challenges, and how they occur, is outside the scope of this current book, they demonstrate the importance of caution in developing a new framework for a ratings and rankings system as this book endeavours to do in ensuring adequate rigor and avoiding overreliance on certain source data.

A relatively new ratings and rankings system, the Mo Ibrahim Foundation's *Ibrahim Index of African Governance*, "measures and monitors governance performance in African countries," (2018). As such, it offers an annual assessment of only African countries across four categories of governance, which it defines as "the provision of the political, social and economic public goods and services that every citizen has the right to expect from his or her state, and that a state has the responsibility to deliver to its citizens," (2018). The four categories of governance include safe and rule of law, participation and human rights, sustainable economic opportunity, and human development, which are further broken down in to 14 sub-categories that include 100 indicators of governance. While the breadth of the indicators included in the *Ibrahim Index of African Governance* is quite extensive, measuring over 100 difference aspects of governance, the system does suffer from some of the same challenges as other ratings and rankings systems in that it relies on existing data sources, which themselves are composite indicators of other indicators. As with the World Bank's WGI, this index relies on the Heritage Foundation, Freedom House, and Transparency International's indices, which are discussed in other sections, in addition to indicators that comprise these systems themselves, such as those used by the Bertelsmann Stiftung, the Economist Intelligence Unit, Varieties of Democracy Institute, and even the World Bank. Despite the index's origination in 2007 as a more recent system compared to the World Bank's WGI or Freedom House's Freedom in the World, it still encounters challenges as it collects and collates data from other sources rather than being an originator of data itself. These challenges are even further complicated as the index itself serves as a data provider to some of its own sources of data, including the Economist's Intelligence Unit, the Varieties of Democracy Institute, and the Centre for Democratic Development Ghana's *African Electoral Index* and *Sanction in Africa Index*.

2.1 Ratings and rankings systems: economics and business

In addition to measures of democracy, human rights, and governance, economic ratings and rankings systems also exist that offer insights in to countries' levels of economic freedom, business environment, labour conditions, and overall prosperity.

The Heritage Foundation's *Index of Economic Freedom* is an example of one such economic rating and ranking system. The Heritage Foundation, a right-leaning think tank based in the United States, measures economic freedom among four categories that each have three sub-categories with indicators scaled on a range of 0 to 100 attached to them. These four categories include the rule of law, which measures property rights, government integrity, and judicial effectiveness; government size, which includes government spending, tax burden, and fiscal health; regulatory efficiency, which includes business freedom, labour freedom, and monetary freedom; and open markets, which includes trade freedom, investment freedom, and financial freedom. Once a numerical score is assigned a country, it is ranked among all others that are measured in order to produce comparable data. Countries are then further grouped in to five colour categories based on their score ranges (for example, dark green for scores 80 to 100, light green for scores 70 to 79.9, yellow for scores 60 to 69.9, orange for scores 50 to 59.9, and red for scores 0 to 49.9.

The World Bank's *Ease of Doing Business Index* is also a well-known ratings and rankings system used to assess economic issues within countries. The *Ease of Doing Business* ranking for a country is calculated relative to other countries included in the assessment through the calculation of a score known as the distance to frontier, whereby this score "benchmarks economies with respect to regulatory best practice, showing the absolute distance to the best performance," (World Bank 2018). There are 41 individual indicators for a which a distance to frontier score is assessed for each country among a total of 10 different categories, including starting a business, dealing with construction permits, getting electricity, registering property, getting credit, protecting minority investors, paying taxes, trading across borders, enforcing contracts, and resolving insolvency. Given that each indicator is based on best practice as observed globally since 2005, different countries can thereby set the best performance standard across various indicators. For example, the *Ease of Doing Business Index 2018* featured New Zealand as having the best practice in the amount of time for starting a business, whereas Japan had set the best practice in terms of cost for getting electricity. Given the use of the distance to frontier calculation, this helps the World Bank to easily identify when reforms takes place within a country since the magnitude of the change in the distance to frontier gap helps to signal when these changes occur. This illustrates a clear example of the use of a ratings and rankings system such as the *Ease of Doing Business Index* in helping the World Bank to identify progress, or the lack thereof, in implementing policy reforms toward a particular objective (i.e. improving the business environment).

Another sizable, yet thematically interesting ratings and rankings system, is the Legatum Institute's *Prosperity Index*. This index measures nine areas which it believes contribute to prosperity, including a country's business environment, economic quality, environment, health, education, safety and security, social capital, personal freedom, and governance. The index differs from other systems as it seeks to advance a specific understanding of the concept of prosperity, which the Legatum Institute (2017) believes is "created by both economic wealth and social wellbeing working together in a relationship where each benefits and advances the other." As such, the index provides measurements on 104 variables attributed to each of the nine areas listed above, and which the organisation has identified, based on its own academic and analytical work, that determines countries' economic performance and social wellbeing. Although the index performs a number of statistical manoeuvres, including weighting of indicators and areas, it too relies on external data sources, similar to many of the other ratings and rankings systems described previously in this book. Again, the same cast of data sources, such as Freedom House and the World Bank's WGI, serve as important contributors across many of the index's 104 variables.

2.2 Select ratings and rankings systems and associated challenges

As has been illustrated with these examples of ratings and rankings systems, some key challenges come to the fore. In addition to those noted already, we will explore in this section a selection of these assessments further as they typify the challenges faced by others.

The World Bank's *Worldwide Governance Indicators* (WGI) prove to be one such system that has been closely studied. One critique toward the WGI focuses on its conceptual definition of governance, a challenge encountered by other ratings and rankings systems explored elsewhere in this book. "Although the concept of governance is widely discussed among policymakers and scholars, there is, as of yet, no strong consensus around a single definition of governance or institutional quality," (Kraay et al. 2010).

In addition, the World Bank's other indicator system, the *Ease of Doing Business Index*, has also been critiqued previously falling victim to contemporary understandings of governance and regulation during the time it was developed without ensuring relevance over time as social science and development theory evolved. In a 2008 review of the World Bank's *Ease of Doing Business Index*, an evaluation found that "[the *Ease of Doing Business Index*] was, in fact, a symptom of the Zeitgeist at the time…the methodology presumes that less regulation is better than more than regulation, in every case, everywhere," (Engle Merry et al. 2015). Because of this neoliberal approach in its first iterations, this index proposed indicators, and as such measured countries, in an attempt to normalise its underlying assumptions toward the subject matter even though those assumptions were unsettled and had not achieved normative international consensus. For example, Engle Merry et al. (2015) point out that the 'employing workers' component of the *Ease of Doing Business Index* had been promoting an unsettled conceptualisation of labour market regulation, which was being disputed by the International Trade Union Confederation and bypassing International Labour Organisation (ILO) conventions.

This example, although later revised and then eliminated as an indicator by the World Bank, represents one of the pitfalls in which ratings and rankings systems can encounter as they seek to normalise underlying assumptions that go in to the definition and application of indicators, thereby having potential real-world policy implications as powerful, decision-making organisations rely on indicators to guide their work.

Another powerful example, and potential challenge posed by the World Bank's system and indicators more generally in promoting a 'study to the test' approach, is highlighted by Sam Scheuth (2016) when he writes about the country of Georgia:

> "In the case of (Caucasus) Georgia, the government carefully targeted its reforms to raise the country's DBI [*Doing Business Index*] ranking, which vaulted from 100 out of 155 in 2006 into the top 20 by 2008 (World Bank and IFC 2006a 2007). This rankings ascension, in turn, was used in an investment promotion campaign designed to dispel the country's profile as a post-Soviet failed state and recast it as a new "frontier market." Concurrently, the volume of foreign direct investment (FDI) inflows to Georgia tripled between 2005 and 2007, as a results of which it was the world's ninth-highest-ranked recipient of FDI as a percentage of GDP (UNCTAD 2008)…the Georgia investment promotion strategy targeted higher DBI rankings by using its indicators as a schema for legal reforms."

The example of Georgia well demonstrates not only how an organisation's adoption and use of an indicator guides decision-making for policymakers within that organisation, but also how indicators have been internalised by those being measured themselves in order to game their policies toward certain outcomes.

Another such example selected from the above discussion of ratings and rankings systems exists in the form of Freedom House's Freedom in the World survey. Similar to the World Bank's *Ease of Doing Business Index* that has the potential to guide investment and policy decisions toward countries, the Freedom in the World survey is specifically identified as part of the decision-making parameters implemented by the Millennium Challenge Corporation (MCC), a United States government bilateral development organisation. Specifically, when identifying which countries are eligible to receive development assistance, the MCC sets "hard hurdles" that must be met in the realms of democratic rights (defined as political rights or civil liberties) and control of corruption. Whereas the control of corruption indicator is established by the World Bank's WGI assessment, the political rights indicator and civil liberties indicator each used by the MCC are established by Freedom House. The rationale from the MCC (2017) for using these Freedom House indicators is as follows: "Requiring that a country pass either the Political Rights or Civil Liberties indicator creates a democratic incentive for countries, recognizes the importance democracy plays in driving poverty-reducing economic growth, and holds MCC accountable to working with the best governed, poorest countries." In essence, the MCC represents a powerful example whereby indicators are used to inform consequential funding and policy decisions. As such, Freedom House, as the developer of such indicators, holds significant power in this regard, despite the *Freedom in the World* survey also suffering from challenges similar to the World Bank's ratings and rankings systems. One such challenge presented by Nils Steiner (2016) arises when he notes that the Freedom in the World results "tend to favour US allies and/or disfavor major antagonists of the US government," which "…raises concerns that closer links between governments and organisations rating third countries might be dangerous in that they can lead to the presence of political bias in the ratings." The implication present in Steiner's study is that, as an organisation based in the United States and funded in part by the United States government, Freedom House ratings proposed by its system are politically biased, without claims as to whether this is intentional or unintentional.

Sarah Sunn Bush further highlights this challenge, particularly when ratings and rankings systems such as the Freedom House *Freedom in the World* Survey come into formal use by governments. In a discussion of the MCC, Sunn Bush (2017) notes, "…the ideological alignment that encourages U.S. policymakers to use the FITW ratings makes them problematic to use for certain functions." This alignment between the indicators and American policymakers was made possible because the Freedom House survey coincided with and reinforced policymakers' ideas about liberal democracy. This is problematic, particularly when used for contemporary policymaking, because if "…the benchmarks that are used to implement conditionality [of MCC development assistance] are biased in favour of U.S. allies, then the stated purpose of the [MCC] program is undermined," (Sunn Bush 2017). Sunn Bush essentially argues that because the Freedom House survey is already in alignment with American ideas of liberal democracy, those countries most likely to not presently conform with these same ideas (i.e. non-American allies) would not be eligible to receive development assistance, despite it being a stated goal by the MCC that it seeks to encourage democratic policy reform. Given this scenario, the ratings and rankings system, having reached a desired and codified level of use by an influential policymaking organisation (i.e. the MCC), is undercut in its overall goal to promote democracy and human rights by the same factor that made it so successful with policymakers in the first place, namely its alignment with prevailing attitudes and conceptions in the United States of what constitutes a liberal democracy.

The above exercise reviewing non-space related ratings and rankings systems and their associated challenges is not meant to undermine the proposed framework contained herein, but rather to guide and inform around the potential pitfalls such a system may encounter once applied and used. This review also assists in situating the proposed framework as it follows from the two historical movements, namely the evidence-based policy and programming movement associated closely with the rise of ratings and rankings systems across disparate fields of research and practice.

2.3 The Space Security Index

Although numerous indicators and ratings and rankings systems exist across a variety of issue areas as explored above, the *Space Security Index* (SSI) remains the only space-related assessment in existence. This is where this book is expected to make its most significant contribution in offering an additional resource for advocates and policymakers alike on a more general basis for understanding the use of outer space by states, not necessarily focused on only security of the space environment. According to Project Ploughshares (2018), a supporter of the SSI, "The *Space Security Index* is the first and only annual, comprehensive and integrated assessment of space security. The project seeks to provide a policy-neutral fact base of trends and developments in space security based on primary, open-source research in an annual report." Project Ploughshares considers space security a politicised issue area, and views SSI as an important fact-based (i.e. evidence-based) tool in informing policy discussions regarding space security challenges and responses. SSI (2018) describes its efforts as "Based on seventeen indicators of space security, it provides background information and in-depth analysis on key trends and developments in the space field."

In terms of its methodology, SSI first defines space security as "the secure and sustainable access to and use of space and freedom from space-based threats" (2018). SSI structures its annual report around four themes, including condition and knowledge of the space environment, access to and use of space by various actors, security of space systems, and outer space governance. Each of these themes includes a number of indicators; however, only three of the total 17 indicators deal exclusively with outer space governance, the primary emphasis of this book as described in later chapters. The methodology for the SSI report relies upon expert surveys and inputs through working groups that ultimately result in "a broad overview of international perceptions of space security," (Space Security Index 2014).

The SSI differs from the previous systems discussed earlier in this book as it does not necessarily rate or rank countries in terms of its subject matter, space security. Rather, the SSI offers an overview of the state of certain topics each year for its assessment, which then highlights important developments that took place during the year that affected the topic noted in the indicator. Therefore, the use of the word indicator within the SSI is not used in the normative understanding of the term as a way to measure change away from a baseline or toward a target, but rather to simply indicate a specific topic for discussion within report on the four broader themes of space environment, access to and use of space by various actors, security of space systems, and outer space governance. For example, the fourth theme contained in the SSI 2017 report, which is an overview of space security in calendar year 2016, begins with its first indicator, Indicator 4.1: National space policies, conducting an overview of national space policy trends generally before addressing country-specific developments that occurred in 2016 (i.e. the United State Air Force's Space command white paper; the European Defence Action Plan; the Chinese white paper on space activities; the thematic topic of policies regarding space resource utilisation; the African space policy and strategy; and, the adoption of space policies by upcoming spacefaring countries). While the information and analysis presented for this indicator and the others included within the SSI prove insightful and relevant to understanding space security, the SSI fails to take a normative approach to behaviour by states in their access and use of outer space, which nominally serves as a goal for other ratings and rankings systems discussed elsewhere in this book, aside from informing and guiding policymaking. However, this is to be expected of the SSI since it makes no claim to the contrary when it clearly articulates its aim to "to improve transparency on space activities and provide a common, comprehensive, objective knowledge base to support the development of dialogue and policies that contribute to the security and sustainability of outer space," (Space Security Index 2017).

There remains a paucity of comprehensive and periodic review of countries' behaviour and operations in space, especially with regard to international space law and established norms. This is not to contend that research on countries' space policies and space activities does not exist; however, it does not exist in the form explored in the first chapter of this book. The SSI provides the closest periodic review, but it neither assesses countries individually, nor does it present a comparison of relevant countries' behaviours according to international space law and norms in order to evaluate whether space activities and policies take place in the spirit of ensuring access and use of outer space by all. The assessment performed by the SSI also does not follow a rigorous and regularly applied methodology. Indeed, SSI's annual reports rely on expert analysis and are then considered in a review workshop; however, this review is done on a discussion and narrative basis against particular indicators, which are largely certain topics comprising broader themes. The indicators included in the SSI are neither applied to all spacefaring countries per se, nor are they used to assess progress from a baseline measure or toward a target.

2.4 Gaps in existing space policy research

Given the SSI's role in informing debate around states' behaviour in outer space rather than setting the terms of the conversation, this illustrates a challenge confronting the international space community. Presently, there remains a lack of a consistent approach to measure countries' adherence and compliance with international law and normative practices with regards to the access and use of outer space. This lack of measurement is owing to various factors, including difficulties in measuring compliance because of a lack of deliverable and specific measures and indicators within existing international instruments, difficulties in measuring compliance because of open-ended interpretations of international instruments and tracking domestication of said instruments by each signatory state, and disagreement over the weighting of different parts of international instruments by various space actors depending from which sector they approach the issue (i.e. civil society, military, government, or industry). For example, civil society would, hypothetically, prioritise transparency and accountability in the access to and use of outer space, whereas governments would prioritise national security and access components and industry would prioritise use and ownership of space resources.

Additionally, the nascent stage of development for international space law, compared to other areas of international law, combined with the relatively few international space advocacy and research groups (again, comparable to other fields) leads to uncertainty and disagreement on how to interpret alternative and competing approaches for access to and use of outer space. Unequal access to information by different groups across civil society, government, and industry also compounds this challenge in being able to successfully and comprehensively measure and monitor states' behaviour regarding outer space. Therefore, given these conditions and the undefined consensus for what constitutes good behaviour versus bad behaviour for the access and use of outer space, this book will seek to make a contribution in setting normative understandings for how states should behave according to a common baseline despite different, and sometimes competing, priorities and interests. This book remains fully cognitive of the normative language used herein and its liability to be confined to only contemporary understandings of the topic without ensuring flexibility for future iterations and developments that may take place in the field, in addition to potentially subscribing to limited understandings of the subject from a Global North, dominant power, and/or spacefaring state's perspective, given how conceptualisations from these groups of how space should be accessed and used form the basis of many present-day space policy debates.

The following chapter therefore seeks to address these concerns by first further defining the problem, discussing the current condition of international space law and relevant advocacy and research groups, and the impetus for suggesting a normative, yet flexible, inclusive, and responsive framework for how states' behaviours regarding outer space can be assessed.

3 INTERNATIONAL SPACE LAW AND NORMS

The criteria used to develop the framework for assessing countries' compliance with international space law and norms will be derived from existing international legally binding instruments and non-binding guidelines and resolutions adopted by the United Nations Committee on the Peaceful Use of Outer Space (UNCOPUOS) and the United Nations General Assembly (UNGA). UNCOPUOS serves as the venue in which international space law is developed, and the United Nations Office for Outer Space Affairs (UNOOSA) serves the committee's secretariat and to operationalise many of the functions agreed by the committee and the UN more generally.

Since 1967, the international community has developed five treaties relating to the access and use of outer space. These are: the Outer Space Treaty of 1967, the Rescue Agreement of 1968, the Liability Convention of 1972, the Registration Convention of 1975, and the Moon Agreement of 1979 that all comprise the binding forms of international law that relate to the access and use of outer space. However, there are only four of these that have received a majority of state signatories whereas the Moon Agreement has only received a relatively small number of signatories. In addition to the treaties, five relevant declarations and sets of legal principles that relate to outer space include the Declaration of Legal Principles of 1963, the Broadcasting Principles of 1982, the Remote Sensing Principles of 1986, the Nuclear Power Sources Principles of 1992, and the Benefits Declaration of 1996. These treaties and sets of principles will be explored further in the section below, which will be followed by a discussion of relevant provisions of UNGA resolutions.

3.1 Differences in binding and non-binding international instruments

International law, comprising binding commitments in which states subject themselves to being governed and bound by rules of agreements, largely exists in the form of treaties as these binding instruments. This differs from non-binding instruments such guidelines or recommendations that do not have the same level of international accountability. The United Nations Educational, Scientific, and Cultural Organization (UNESCO 2017) summarises this well when it writes of binding instruments versus non-binding instruments: "Binding instruments, or 'hard law', establish rules expressly recognized by the contracting States (Article 38 (1) of the Statute of the International Court of Justice)…By ratifying the instrument, States explicitly recognize their obligation to respect the terms of the treaty." This affects domestic laws of countries as well because "In accordance with the principle of primacy of the international law over national law, State Parties are bound to adapt their national legislation to the provisions of the treaty and introduce all relevant measures in their national legal system to implement their obligations…" (UNESCO 2017).

Separately, according to UNESCO (2017), "Non-binding instruments, or 'soft law', provide guidelines of conduct, which are neither strictly biding norms of law, nor completely irrelevant political maxims…Main examples of non-binding instruments are declarations, recommendations and resolutions." Since international space law and norms established around the access to and use of outer space are found in both binding, international legal agreements (i.e. treaties) and non-binding instruments, it remains important to understand the basis of these non-binding instruments in addition to the binding treaties.

"Declarations do no create legal obligations for States that adopt them. They reflect principles on which these States agree at the time of their adoption and proclaim [as] standards [of behaviour], which though non-binding, [nevertheless] impose moral obligations…recommendations are intended to influence the development of national laws and practices…[and] resolutions are formal expressions of opinion by a legislative body or a public meeting. The resolutions made by the United Nations General Assembly…are the therefore an expression of the Member States of these Organizations," (UNESCO 2017).

3.2 Binding international space law treaties

In this section, this book will explore the binding instruments that comprise international space law in the form of the five outer space treaties. The Outer Space Treaty of 1967, formally known as the Treaty on Principles Governing the Activities of States in the Exploration and Use of Outer Space, including the Moon and Other Celestial Bodies, signifies the first binding instrument used by the international community regarding access to and the use of outer space. In its first article, Article I, it establishes key principles on the exploration and use of space for the benefit of all countries, the freedom of access and exploration of the moon and celestial bodies on a basis of equality, and the freedom of scientific investigation. The treaty goes on to discuss and establish other important principles and key foundational components of international space law, including non-appropriation of celestial bodies through sovereignty claims or by means of use or occupation, peaceful use of space and the non-placement of nuclear weapons in space, provision of assistance to astronauts during emergencies, irrespective of their origin, state responsibility and authorisation and supervision for national activities in space, liability of launched objects, registration of launched objects, and reciprocity of use of space assets on celestial bodies. Indeed, as seen in upcoming sections, some of these provisions established in the Outer Space Treaty formed the basis of the other subsequently adopted binding instruments, including the Rescue Agreement, the Liability Convention, and the Registration Convention. The Outer Space Treaty, in addition to the others mentioned here, will form the basis of the criteria comprising the framework described later in this book, accompanied by criteria also assigned to non-binding documents regarding outer space.

Following the Outer Space Treaty, the Rescue Agreement, formally known as the Agreement on the Rescue of Astronauts, the Return of Astronauts and Return of Objects Launched into Outer Space, was adopted in 1968. Article I of the Rescue Agreement clearly lays out the responsibilities of states pertaining to accidents and emergencies involving astronauts: "Each Contracting Party which receives information or discovers that the personnel of a spacecraft have suffered accident or are experiencing conditions of distress or have made an emergency landing or unintended landing in territory under its jurisdiction…shall immediately notify the launching authority or…immediately make a public announcement by all appropriate means of communication," (UNOOSA 2017). Article 2 specifically states that "…a Contracting Party [state]…shall immediately take all possible steps to rescue them [astronauts] and render them all necessary assistance," (UNOOSA 2017). Here is where the Rescue Agreement gains its name as it establishes the principle of rescue and assistance for all astronauts, irrespective of origin or launching state, in the case of emergencies. In subsequent articles of the treaty, it goes on to establish principles for providing assistance if states are able to do so despite distressed astronauts not being under their jurisdiction, the safe and prompt return of astronauts to their launch state once found, and the notification, recovery, and return of space objects and component parts.

The third international treaty on outer space is the Convention on International Liability for Damage Caused by Space Objects, commonly known as the Liability Convention. The convention begins by noting its assumption that space actors intend to operate safely when they launch space objects "…taking into consideration that, notwithstanding the precautionary measures to be taken by States and international intergovernmental organizations involved in the launch of space objects," (UNOOSA 2017). This suggests that the treaty assumes space actors act in good faith with regard to space, and in particular the launching of space objects, which may increasingly prove to be a challenging assumption to make as space activities increase and the domain becomes increasingly congested and contested. Information and analysis confirming, or denying, this assumption and others like it that form the basis of international space law remains one of the areas in which this book strives to contribute to informed decision-making and policymaking. Nonetheless, similar to those before it, the Liability Convention establishes a number of principles which states must follow to remain compliant with international law. These principles, created in its several articles and provisions, include: the absolute liability of launching states for damage caused by their space objects to anything on the Earth's surface or flying aircraft; the process of establishing fault for damage done to other space objects; the responsibility and payment of compensation to third-party states if damage results from two launching states; the joint responsibility and liability of launching states if they jointly launch a space object; the exoneration of liability of launching states due to gross negligence or act or omission with intent to cause damage by the state claiming compensation; the claims process for compensation; the statute of limitations of one year for a claim to be made following the date of occurrence of the damage or the identification of the liable launching state; and the provision of appropriate and rapid assistance to a damage-affected state in the case of larger-scale danger to human life or interference to living conditions or functioning of important centres.

Given the emphasis on compensation for damage caused by launching states to the Earth's surface, aircraft, and space objects with varying levels of liability depending on where the damage occurs, some have argued that the Liability Convention promotes the larger international space law norm prohibiting the militarisation of space, and therefore the conduct of damage-causing space warfighting activities. Pavle Kilibarda (2017) describes this well in writing, "Whereas militarization in the broadest sense is legal, the concept of liability at least constrains the 'weaponization of space' (means as the use of outer space for direct force deployment in situations of armed violence)…As a matter of treaty interpretation, it seems absurd to suggest that a treaty which clearly covers cases of accident would not also apply to damage caused deliberately, or is limited to deliberate damage caused outside of an armed conflict." The interpretation of the Liability Convention alongside other documents comprising international space law assists with determining how military space activities should be treated in the subsequent chapters of this book.

The fourth, and remaining commonly adopted, international treaty on outer space is the Registration Convention, formally known as the Convention on Registration of Objects Launched into Outer Space. This treaty helps to further explain and codify elements established in previous treaties regarding the registration of space objects. The Registration Convention, opened for ratification in 1975. In Article II, it makes it the responsibility of launching states to develop and maintain individual, national registries of objects they launch into space and to register on an ongoing basis these objects in the registry. In subsequent articles of the convention, it goes on to create other, more specific provisions, including: an instruction to the UN Secretary-General for the UN to maintain its own registry of space objects based on information provided to it by launching states. The information to be provided includes the details of the launch and characteristics of the space object that was launched, which should also be included in the individual launching states' and UN registries. The convention also details the process for how states should identify other space objects that cause damage to their own spacecraft.

The fifth and final treaty regarding the access to and use of outer space is the Moon Agreement, otherwise known as the Agreement Governing the Activities of States on the Moon and Other Celestial Bodies. The Moon Agreement is the least popular outer space treaty, having had only 18 states that have ratified or acceded to it. This may be due to the treaty featuring a number of restrictive terms and provisions, and the introduction of a principle in Article 4 stating "The exploration and use of the Moon shall be the province of all mankind and shall be carried out for the benefit and in the interests of all countries…" (UNOOSA 2017). This is later elaborated in Article 11 as well when it notes "The Moon and its natural resources are the common heritage of mankind…" (UNOOSA 2017). Other restrictive provisions included in Article 3 forbid military bases, installations, and fortifications; weapons testing; and military manoeuvres on the Moon; the placement of weapons of mass destruction in orbit, on, or in the Moon; whereas Article 4 goes on to also state "Due regard shall be paid to the interests of present and future generations as well as to the need to promote higher standards of living and conditions of economic and social progress and development" (UNOOSA 2017) regarding the use of the Moon. Perhaps the most contentious of the Moon Agreement's provisions regards its position of the use of the Moon's natural resources established through the aforementioned common heritage of mankind principle it sets forth:

> "States Parties to this Agreement hereby undertake to establish an integrational regime, including appropriate procedures, to govern the exploitation of the natural resources of the Moon…The main purposes of the international regime to be established shall include…an equitable sharing by all States Parties in the benefits derived from those resources, whereby the interests and needs of the developing countries, as well as the efforts of those countries which have contributed either directly or indirectly to the exploration of the Moon, shall be given special consideration." (UNOOSA 2017).

Given such emphasis on resource sharing, leading spacefaring countries have not ratified or acceded to the Moon Agreement, making it essentially an ineffective and, in practice, negligible treaty with no effect on space activities. Whereas the treaty aims to establish a number of best practices and guiding principles for the cooperative use and exploitation of the Moon and celestial bodies, particularly from a Global South perspective with its emphasis on the distribution of benefits to developing countries who may otherwise lack access to such resources and the ability to benefit from them, it fails to meet the necessary geopolitical and economic realities of the current space environment and the priorities of those engaging in activities in the domain. Thus, given the Moon Agreement's low level of ratification and accession by countries and its minimal effect on current outer space governance, it will not be considered for purposes of developing the framework proposed in later chapters of this book because it does not reflect the interests of spacefaring countries and the standards to which they hold each other accountable.

3.3 Binding international space-related treaties

Now that the main five treaties on outer space have been discussed in the previous section, it remains important to take note of the other space-related treaties that will be considered for purposes of assessment within the framework for understanding countries' level of compliance with international space law and norms. These specific treaties will not be discussed at detail in the current section as was done for the previous five treaties as these do not deal exclusively with outer space and may prove too institution, technical, or project-specific, making them not useful for interpreting and arriving at larger and more applicable norms of behaviour in outer space. An assessment will be made in the upcoming chapters for whether provisions of these treaties are incorporated in the framework's criteria; however, for purposes of this section, it is instructive to note what these space-related treaties are in order to acknowledge the context in which this book seeks to make its contribution. Therefore, these space-related treaties include the Treaty Banning Nuclear Weapons Tests in the Atmosphere, in Outer Space, and under Water of 1963; the Convention Relating to the Distribution of Programme-Carrying Signals Transmitted by Satellite of 1974; the Agreement Relating to the International Telecommunications Satellite Organization of 1971; the Agreement on the Establishment of the INTERSPUTNIK International System and Organization of Space Communications of 1971; the Convention for the Establishment of a European Space Agency of 1975; the Agreement of the Arab Corporation for Space Communications of 1976; the Agreement on Cooperation in the Exploration and Use of Outer Space for Peaceful Purposes of 1976; the Convention on the International Mobile Satellite Organization of 1976; the Convention Establishing the European Telecommunications Satellite Organization of 1982; the Convention for the Establishment of European Organization for the Exploitation of Meteorological Satellites of 1983; and the International Telecommunication Union Constitution and Convention of 1992.

3.4 Non-binding international principles, resolutions, and guidelines

In this section, we discuss the various principles, resolutions, and guidelines adopted by the United Nations General Assembly (UNGA). Principles adopted by the UNGA include the Declaration of Legal Principles Governing the Activities of States in the Exploration and Use of Outer Space (1963), Principles Governing the Use by States of Artificial Earth Satellites for International Direct Television Broadcasting (1982), Principles Relating to Remote Sensing of the Earth from Outer Space (1986), Principles Relevant to the Use of Nuclear Power Sources in Outer Space (1992), and the Declaration on International Cooperation in the Exploration and Use of Outer Space for the Benefit and in the Interest of All States, Taking into Particular Account the Needs of Developing Countries (1996). Furthermore, UNGA-adopted resolutions include the 1961 resolution on international cooperation in the peaceful uses of outer space, the 2000 resolution on the use of geostationary orbit, the 2004 resolution on the concept of the launching state, the 2007 resolution on the registration of space objects, and the 2013 resolution on recommendations for national legislation relevant to the peaceful exploration and use of outer space. Finally, UN guidelines and frameworks developed to guide the use of outer space include the 2007 Space Debris Mitigation Guidelines of the Committee on the Peaceful Uses of Outer Space, and the 2009 Safety Framework for Nuclear Power Source Applications in Outer Space.

The Declaration of Legal Principles Governing the Activities of States in the Exploration and Use of Outer Space (1963) represented the first UNGA-level legal expression for how states should operate in outer space. This later evolved into the Outer Space of Treaty of 1967, which represented the culmination of these non-binding principles in a binding international treaty which states could ratify. As was later included in the Outer Space Treaty, the Declaration of Legal Principles by which states should abide include: the exploration and use of outer space for the benefit and in the interests of all mankind; freedom of exploration of space and celestial bodies; non-appropriation of space and celestial bodies; state responsibility and authorization and supervision of space activities; non-interference in space; registration and liability of launched space objects; and the provision of assistance to astronauts regardless of nationality. For purposes of the framework that is later discussed in this book, the Declaration of Legal Principles will not be included in the rating system as the same concepts discussed in it are revisited in the Outer Space Treaty, which represents the binding international law form of these same principles. Therefore, the Outer Space Treaty will be included among the rating criteria, whereas its predecessor, the Declaration of Legal Principles, will not.

In addition to the Declaration of Legal Principles, the second non-binding set of principles adopted by the UNGA include the Principles Governing the Use by States of Artificial Earth Satellites for International Direct Television Broadcasting (1982). This set of principles was adopted by the UNGA to reaffirm "…international direct television broadcasting by satellite should be carried out in a manner compatible with the sovereign rights of States, including the principle of non-intervention, as well as with the right of everyone to seek, receive, and impart information and ideas…" (UNOOSA 2017). The principles also establish the principles of free dissemination and mutual exchange of cultural and scientific information and knowledge, contributions toward educational, social, and economic development, equal rights of all states to broadcast by satellites, international cooperation and special assistance to developing countries for international broadcasting, settlement for disputes, state responsibility, duty and right to consult between the broadcasting state and receiving state, protection of copyright and neighbouring rights, and notification to the United Nations.

Following the UNGA's adoption of the principles relating to international broadcasting, the UNGA adopted the Principles Relating to Remote Sensing of the Earth from Outer Space in 1986. Similar to many previously discussed international instruments regarding outer space, this set of principles begins by affirming that remote sensing activities should be carried out for the benefit and in the interest of all countries, in addition to defining what is meant by remote sensing, primary data, processed data, analysed information, and remote sensing activities. Interestingly, when defining remote sensing, the UNGA limits its definition to only those activities that are used for "the purposes of improving natural resources management, land use and the protection of the environment," (UNOOSA 2017) which only captures a portion of the scope of remote sensing activities given their applicability in other contexts, notably defence and intelligence. The principles go on to further outline that remote sensing should not be conducted in a manner detrimental to the legitimate rights and interests of the states being sensed; remote sensing states should include other states in their remote sensing activities; remote sensing states should establish and operate data collection and storage and processing and interpretation through agreements with other states; states should provide technical assistance to other states on remote sensing activities, notification of the United Nations of remote sensing activities; remote sensing activities should promote the protection of the environment and natural resources; states that have remote sensing activities should furnish information if it is capable of preventing harm to the environment and mitigating the effect of natural disasters; and the rights of sensed states to have access to primary, processed, and analysed data for territory under their jurisdiction on a non-discriminatory basis and for a reasonable cost.

Another set of principles the UNGA adopted include the Principles Relevant to the Use of Nuclear Power Sources in Outer Space, which were adopted in 1992. The first unique principle established in this set features the principle to minimise the amount of radioactive material in space by restricting its use to only those missions that could not be operated in a reasonable way by non-nuclear sources, and it is noted as such in the context that states should exercise as much caution as possible and with high a degree of confidence that nuclear material should not affect individuals, populations, or the biosphere with contamination. These principles go on to note numerous items, including those that reference previously established standards by other international authorities such as the International Commission on Radiological Protection. These principles are: the protection of the public against radiation according to internationally set standards; limited exposure during accidents through the proactive design and construction of nuclear power systems that conform to international standards; the incorporation of the defence-in-depth concept during design, construction, and operations of nuclear systems; the operation of nuclear reactors only on interplanetary missions, sufficiently high orbits, or low-Earth orbits so long as post-mission they are disposed of in high orbits; the development of nuclear reactors using only uranium-235 fuel and operations only once in orbit; the conduct of a safety assessment prior to launch; notification of re-entry of a nuclear power source to those states that may be affected; consultation with concerned states; assistance for tracking of a re-entry of a nuclear power source, and assistance by the launching state and others with the capabilities to eliminate the effects of a re-entered object within those states that are affected; the responsibility of states for objects they or those within their jurisdiction launch; and liability and compensation for damage caused in accordance with existing international law.

The final, and most recent set of principles adopted by the UNGA, include the Declaration on International Cooperation in the Exploration and Use of Outer Space for the Benefit and in the Interest of All States, Taking into Particular Account the Needs of Developing Countries (the so-called Benefits Declaration) which was adopted in 1996. These principles essentially affirm other sentiments expressed elsewhere in international instruments on outer space regarding the participation and mutual benefit of all states in the exploration and use of outer space. In particular, the principles note: the free determination of all states regarding their participation with other states on an equitable and mutually acceptable basis; the fairness and reasonableness of cooperative contracts between states; the contribution of leading spacefaring countries to new and upcoming space programmes by developing countries through international cooperation activities; the use of various means of cooperation including government-to-government, non-government-to-non-government and global, multilateral, regional, and bilateral; the promotion and development of space science and technology; the exchange of expertise and technology between states; the use of space applications for developing states to achieve their development goals with support from various organisations and agencies; and the strengthening of UNCOPUOS as an international forum information exchange and the further developing of the UN Programme on Space Applications by countries with the requisite capabilities.

Separate from UNGA-adopted principles as described above, the UNGA has adopted other resolutions related to the exploration and use of outer space. The first of these related resolutions was the 1961 resolution on international cooperation in the peaceful uses of outer space, which predates both the 1963 Declaration of Legal Principles and 1967 Outer Space

Treaty. Given that constitutes the earliest expression of a UN position on the exploration and use of outer space, its relevance can be found in the provisions contained within the Declaration on Legal Principles and the Outer Space Treaty, which are inherited, albeit in a more detailed expression, from those contained within the 1961 resolution. Similarly, the 1961 resolution encourages the registration of launches with UNCOPUOS; tasks the UN Secretary-General with maintaining a registry of launches; and charges the UN Committee on the Peaceful Uses of Outer Space with maintaining contact with governments on outer space affairs, providing the exchange of information, and promoting international cooperation in space.

The UNGA also affirmed in 2000 its support of the UNCOPUOS Legal Subcommittee when it endorsed the subcommittee's agreement on the use of the geostationary orbit as a finite and limited natural resource. Essentially, the Legal Subcommittee of UNCOPUOS via the UNGA recommended that: states should coordinate on access to and use of geostationary orbits in an equitable manner according to the International Telecommunication Union's (ITU) rules; states already with access to geostationary orbit and its spectrum should take steps to ensure developing countries seeking access to the same resource should have equitable access; and states, including developed and developing, should file their requests for access to the orbit via the ITU. As described earlier in the agreement made by the Legal Subcommittee, these recommendations were made with regard to the previously existing practice of "first come, first served" concerning geostationary orbit, which limited in practice access to the orbit by developing countries since developed countries had already been active there given their advanced capabilities in outer space.

Similar to the 2000 resolution, the UNGA adopted a resolution in 2004 on the application of the concept of the launching state. Contained in the resolution, the UNGA made a number of recommendations for which states should consider for conduct of their space activities, including best practices for codifying international space law. The resolution specifically noted that states should enact and implement national laws that allowed for the authorization and supervision of activities in outer space of non-governmental entities within their jurisdiction; states should implement conclusions to their joint launch and cooperation programmes with regard to the Liability Convention; UNCOPUOS should request information from states on their current practices of on-orbit transfer of ownership of space objects; states should harmonise their domestic and national laws with international law; and UNCOPUOS should assist states in developing national laws by providing them with information on the outer space treaties.

The 2004 resolution, which specifically highlighted the development of national laws in alignment with international law, was further elaborated upon in a 2007 UNGA-adopted resolution that included recommendations on enhancing the practices of states and international intergovernmental organisations in registering space objects. Specifically, the 2007 resolution encouraged states and international intergovernmental organisations to ratify or accede and follow practices set out in the binding Registration Convention, in addition to encouraging states to provide uniform and expanded information on their space objects. Moreover, the resolution acknowledged the existing fact that some states and international intergovernmental organisations had yet to ratify and follow the Registration Convention, and therefore suggested a solution should be sought to deal with how space objects of these groups should be registered even though their status was still unagreed. Interestingly, UNGA resolution 1721 B (XVI) adopted in 1961 also serves as an alternative avenue for states to register space objects in addition to the Registration Convention; however, the lack of registration of certain states and international intergovernmental organisations, either through the Registration Convention or the 1961 resolution, still remains problematic. The 2007 resolution also continued to build off of the 2004 resolution regarding the transfer of ownership of space objects, and instructed states on the types of information that should be provided to the UN regarding such transfers, in addition to requesting UNOOSA to provide a model registration form including the information states should supply for registration purposes and means in which states could contact focal points concerning the space objects registered by each state.

As with the 2004 and 2007 resolutions, the UNGA's 2013 resolution included even more detailed recommendations on national legislation relevant to the peaceful exploration and use of outer space. This resolution specifically outlined which types of activities national laws and frameworks should include regarding best practice and compliance with international laws. The UNGA's recommendations suggest that states' regulatory frameworks should account for: the launch of objects into and their return from outer space, the launch and re-entry sites, operations, and control of space objects; the design and manufacture of spacecraft; the application of space science and technology and exploration activities; and determination of jurisdiction of space activities and the authorisation and supervision of such activities. Specifically regarding authorisation, the resolution established that states should set the conditions and procedures for different steps in the authorisation process, including granting, modifying, or suspending and revoking authorisation, and making these steps clear in legal and regulatory frameworks. On supervision, the UNGA recommended states should conduct on-site inspections and require general reporting by space actors, and that administrative procedures and penalties should be in place should space actors not be in compliance with the licensing regime. Other recommendations in the resolution include suggesting to states that the Space Debris Mitigation Guidelines, discussed elsewhere in this book, are followed by those conducting space activities, in addition to prescriptions on how states should maintain their national registry of space objects and how states should ensure processes for claiming and receiving compensation from damage caused by space objects and how states should record and report the transfer of ownership of space objects already in orbit.

In addition to the above principles and resolutions adopted by the UNGA, it has also endorsed, in coordination with work done through UNCOPUOUS, non-binding guidelines for states to follow regarding their activities in outer space. Specifically, these include the Space Debris Mitigation Guidelines and the Safety Framework for Nuclear Power Source Applications. In recognition of the growing problem and threat posed by space debris, the UNGA endorsed in 2007 seven guidelines proposed by UNCOPUOS. These guidelines are as follows:

1) Limit debris released during normal operations;

2) Minimize the potential for break-ups during operational phases;

3) Limit the probability of accidental collision in orbit;

4) Avoid intentional destruction and other harmful activities;

5) Minimize potential for post-mission break-ups resulting from stored energy;

6) Limit the long-term presence of spacecraft and launch vehicle orbital stages in the low-Earth orbit (LEO) region after the end of their mission; and

7) Limited the long-term interference of spacecraft and launch vehicle orbital states with the geosynchronous Earth orbit (GEO) region after the end of their mission.

Each of these guidelines is elaborated in the resolution document; however, for purposes of this current section, they are listed here and will be detailed later in this book when used to help inform the development of criteria comprising the compliance assessment framework.

UNCOPUOS guidelines contained with the Safety Framework for Nuclear Power Source Applications in Outer Space also provide helpful areas in which to measure compliance with best practices of space activities. The Safety Framework, endorsed in 2009 by UNCOPUOS, has an overriding safety objective similar to that contained in the Principles Relevant to the Use of Nuclear Power Sources in Outer Space, to "protect people and the environment in Earth's biosphere from potential hazards associated with relevant launch, operation and end-of-service phases of space nuclear power source applications," (UNOOSA 2017). Beyond this, the guidelines go on to suggest specific issue areas that states' governments must address when considering the use of nuclear power sources (NPS) in space activities, and makes inter alia following recommendations:

1) "Governments that authorize or approve space nuclear power source missions should establish safety policies, requirements, and processes;

2) The governments mission approval process should verify that the rationale for using the space nuclear power source application has been appropriately justified;

3) A mission launch authorisation process for space nuclear power source applications should be established and sustained; and

4) Preparations should be made to respond to emergencies involving a space nuclear power source," (UNOOSA 2017).

Separate from its prescriptions toward states' governments, the framework acknowledges that responsibilities should also rest with the management of organisations deploying nuclear power in space activities, and therefore the framework makes the following recommendations targeted toward these managers and their organisations:

1) "The prime responsibility for safety should rest with the organisation that conducts the space nuclear power source mission;

2) Effective leadership and management for safety should be established and sustained in the organisation that conducts the space nuclear power source mission;

3) Technical competence in nuclear safety should be established and maintained for space nuclear power source applications;

4) Design and development processes should provide the higher level of safety that can reasonably be achieved;

5) Risk assessments should be conducted to characterise the radiation risks to people and the environment; and

6) All practical efforts should be made to mitigate the consequences of potential accidents," (UNOOSA 2017).

Indeed, as was done in the Space Debris Mitigation Guidelines, the provisions within this framework are further detailed in the UNCOPUOS resolution; however, for purposes of this current section, the NPS guidelines are only described in brief and will be further detailed in later sections as they are considered for specific criteria require for developing the compliance assessment framework.

Beyond the Space Debris Mitigation Guidelines and the Nuclear Power Source Safety Framework, a set of voluntary guidelines also requires attention and inclusion within the framework proposed in the next chapter. These guidelines are the recommendations from the Group of Governmental Experts (GGE) on transparency and confidence-building measures (TCBM), the European Union's International Code of Conduct on Space Activities, and the UNCOPUOS guidelines on the long-term sustainability (LTS) of outer space. The guidelines contained within these three separate initiatives are different from past initiatives because they are driven from the "bottom-up" and have involved in some cases input from non-state actors (Secure World Foundation 2018). For example, Russia proposed the GGE, which was adopted by the UNGA in 2010 and eventually convened in 2011 to include experts from 15 different countries who made "recommendations on how governments can share information with an aim to creative mutual understanding and trust, reducing misperceptions and miscalculations and thereby helping to prevent military confrontation and to foster regional and global security," (Secure World Foundation 2018). Similarly in 2010, the European Union proposed its Code of Conduct to promote international best practices for space activities; however, this initiative has stalled since 2015 because of disagreements as to the legal status of the Code of

Conduct and whether countries would see it in competition with legally binding space and arms control treaties (Secure World Foundation 2018). Nonetheless, the European initiative reflected a good faith effort, as did the Russian GGE initiative, to outline and create the basis for normative behaviours that countries could follow in carrying out their space activities.

Perhaps the broadest and most subscribed to initiative of the three described here is the set of guidelines proposed by the UNCOPUOS Working Group on the Long-Term Sustainability of Outer Space Activities. As with the other two initiatives, the UNCOPUOS working group formed in 2010 and concluded its work in June 2018, which resulted in 21 guidelines agreed to by the UNCOPUOS member states through its consensus-based decision-making process. In its role as a convening space for emerging, intermediate, and advanced spacefaring countries, UNCOPUOS's LTS guidelines are meant to help develop national and international practices and frameworks while also remaining flexible enough to adapt the frameworks to country-specific circumstances and conditions (Secure World Foundation 2018). Specifically, the LTS guidelines are intended for supporting international organizations and countries to develop policies that "avoid causing harm to the outer space environment and the safety of space operations," (Secure World Foundation 2018). Importantly, the LTS guidelines cover a broad variety of space sustainability topics and are organised according to four categories: policy and regulatory framework for space activities; safety of space operations; international cooperation, capacity-building and awareness; and scientific and technical research and development. The guidelines range from high-level recommendations such as Guideline A.1 "Adopt, revise and amend, as necessary, national regulatory frameworks for outer space activities" to specific, granular-level recommendations such as a Guideline B.4 "Perform conjunction assessment during all orbital phases of controlled flight," (Secure World Foundation 2018). The set of LTS guidelines, along with the European Unions' Code of Conduct and the GGE's transparency and confidence-building measures, are incorporated in the assessment framework as an important consideration for countries to follow in terms of non-binding international norms of behaviour in outer space.

Given this book has now discussed the two movements for evidence-based programming and policymaking alongside a review of existing ratings and rankings systems and consideration of their potential challenges, in addition to a review of existing binding and non-binding international space law and established norms via the UN system, the upcoming chapters will describe a framework, and the criteria contained therein, for assessing countries' levels of compliance and adherence to international space law and norms. This proposed framework will be placed within the growing world of other ratings and rankings systems, albeit with an emphasis on space and fulfilling an existing gap in space policy literature whereby an annual and systematic, open-source assessment of countries' space policies and behaviours does not yet exist.

4 ELEMENTS COMPRISING ASSESSMENT METHODOLOGIES

Now that this dissertation's contribution has been placed within existing literature and movements, in addition to having reviewed the subject matter of international space law and UN established norms, this chapter will first discuss the methodological elements that will guide the development of a framework and associated criteria before moving to a discussion of the actual framework and structure with which an assessment of countries' space policies can be conducted. For this section discussing and identifying the key elements which should be considered for a ratings and rankings system that relies upon best practices, the following questions will be considered as these are derived from a 2014 study performed by Rachel Gisselquist (2014) based upon her "review of the research literature and on three years of research in practices, specifically the author's experience in developing a well-used measure of governance [i.e. the Ibrahim Index of African Governance]." Specifically, Gisselquist's set of guiding questions will be used to develop this assessment framework because they help to address issue areas that she and others have identified as problematic as it pertains to such assessments, including needing to attend to social science methodology fundamentals, such as concept formation, content validity, reliability, replicability, robustness, and relevance; descriptive complexity, theoretical fit, the precision of estimates, and correct weighting. The ten guiding questions that will be addressed in the upcoming sections are:

1. What precisely does it [the assessment] aim to measure?

2. Does the operational definition capture the concept?

3. How good (reliable, valid, and complete) are the data?

4. Is the measure (including all of its sub-components) transparent and replicable?

5. How sensitive and robust is the measure to different data and design choices?

6. Does the measure allow the analyst to address key questions of interest?

7. Does the measure fully capture governance in all its complexity?

8. Does the measure behave as theory predicts?

9. How precise are index values and are confidence intervals specified?

10. Is the weighting correct?

Prior to answering each of these questions as it pertains to the proposed assessment, Gisselquist (2014) also proposed three additional issue areas that should be considered for those producing such indices:

> "First, in deciding whether to produce a new governance index, they should consider its value-added in a field with dozens of existing measures. Second, does the utility of the index justify its costs?... Third, legitimacy: Governance assessments can have major real-world implications, from aid allocations to investor perceptions. Will the index as designed and implemented be considered legitimate by those assessed?"

Regarding the first question proposed above, the proposed assessment does add value as it is making a contribution in a niche field whereby no annual and systematic reviews of countries' compliance with international space law and norms exists. As discussed previously, the *Space Security Index* (SSI) comes closest to offering this type of review; however, its indicators are largely topic and issue areas that are used for guiding and structuring the annual report rather than quantitative indicators in the evaluative definition of the word in terms of defining a baseline and measuring change against that baseline. The indicators within the SSI are used to indicate a specific topic for discussion within report on the four broader themes of space environment, access to and use of space by various actors, security of space systems, and outer space governance. Additionally, only one indicator, placed under the theme of outer space governance as defined by the report, specifically addresses national space policies. Therefore, this suggests that the SSI may prove to be useful a resource from which to draw data in the application of this dissertation's proposed assessment framework; however, as a report itself, it does not fall within the realm of an international rating and ranking system.

The second question posed above concerns costs of applying the methodology through research. As this framework is novel in that this is the first time it is being proposed, it remains difficult to answer question since there is no past evidence from which to draw. However, it can be projected that the research behind the assessment would rely upon existing data, and in the cases where they there may gaps, researchers would conduct a desk review of national laws and policies and perform media analysis of space activities conducted by countries. This suggests time and effort spent by researchers would come at a cost for implementing the assessment, although other costs, such as travel or access fees, would likely prove minimal given many of the documents needed for the review sit in the public domain. The context of an increasingly congested and contested outer space domain accompanied by rise in new and challenging issue areas, such as space debris and space traffic management, justifies the need for a new tool to assess compliance with international laws and norms, particularly given no such tool exists as yet and the estimated costs for implementing it would only include time and effort of researchers accessing and analysing public domain and open source information.

The third and final question that must be addressed, prior to following the ten guiding questions described previously, concerns whether this assessment will be considered legitimate by those it is assessing and by others using the assessment. Given the data used to inform the analysis of the assessment will be derived from public domain and open access UN repositories and countries' own documents, the legitimacy of the data should not be a concern. Moreover, the standard of assessment used in the framework itself will be based on international treaties, resolutions, principles, and guidelines adopted at the UN level and for which the majority of the international community have endorsed. For example, while the Moon Agreement may exist as a binding international legal instrument governing space activities on the Moon and other celestial bodies, it will not be included in this assessment for purposes of defining which criteria countries should be assessed since very few countries have ratified or acceded to that treaty. The Moon Agreement has been ratified by so few countries (11 signatories; 18 parties as of 2019), and none with a history of lunar exploration or imminent plans to explore the moon or use its resources, that it would not be instructive to apply the framework of this work to assessing implementation practice of countries regarding the Moon Agreement. It would thus be unfair to measure countries according to a set of rules they themselves have not ratified.

Important distinctions such as this should help to alleviate concerns regarding the legitimacy of the framework, in addition to the following ten considerations that will be made as it is developed since these ten guiding questions follow best practices for international ratings and rankings systems. Given these preliminary questions have been addressed situating the proposed assessment in terms of its value-added, justification, and legitimacy, it now remains to discuss the framework with regard to the ten guiding questions asked by Gisselquist.

4.1 What precisely does it aim to measure?

The assessment proposed herein aims to measure how well countries comply with international space law and norms as set forth in the previously discussed international treaties, resolutions, and guidelines. For purposes of defining compliance, this concerns how countries adhere to, follow, and observe these laws and norms both in theory (i.e. existing domestic laws, policies and regulations, and/or frameworks) and in practice (i.e. the effective regulation of space activities that countries conduct). Conceptually, the international community has yet to define many issues regarding the exploration and use of outer space. For example, the demarcation point between airspace and outer space remains an unsettled and open issue in air and space law (Oduntan 2003). However, existing disagreements, and therefore the accompanying lack of conceptual clarity, do not hinder the conceptual strength upon which this framework will rely as the assessment is primarily utilitarian and measuring what already exists and has been agreed upon by the international community. In fact, this framework will purposely exclude areas in which agreement and conceptual clarity have not been achieved in order to ensure its legitimacy and relevance among space actors.

4.2 Does the operational definition capture the concept?

Gisselquist (2014) argues that "Once a concept has been properly specified, the next step logically and chronologically, is to operationalize it. An operational definition should identify the component(s) to be included in the measure and specify how these components are put together in a manner that is consistent with the core concept." As has been done in earlier sections defining what constitutes international space law and norms, this question has been answered. Indeed, the components comprising international space law and norms have been identified as comprising both the binding outer space treaties and other space-related treaties and agreements, and the non-binding UN principles, resolutions, and guidelines. These documents constitute, for purposes of this assessment, international space law and norms, and are therefore valid in terms of the measurement's content.

4.3 How good (reliable, valid, and complete) are the data?

This third question is posed to move beyond only measuring the validity of the assessment's concept and definitions by also forcing the consideration of data quality and its validity and reliability. "Validity refers to whether the measure accurately captures what it purports to capture. Reliability refers to the consistency of the measure," (Gisselquist 2014). The data used for this assessment will primarily be public domain and open source documents developed by the UN and countries themselves. UN-produced data is expected to likely be valid and reliable; however, country-produced data, and the availability of data country to country, may prove challenging. Indeed, countries may have in place domestic laws and regulations that should exist in the public domain, but which may be unpublished or difficult to access. Additionally, some countries may not have laws and regulations in place in which to measure compliance, or may not conduct space activities, thereby limiting the pool of countries that may be assessed. Therefore, this assessment will initially strive to only measure those countries with sufficiently valid, reliable, and complete data in which to apply the framework. This consequently implies that only spacefaring countries will at first be assessed, but this pool of countries will likely grow as more entrants appear in the outer space domain and as more governments seek to participate in space policymaking fora.

4.4 Is the measure (including all of its sub-components) transparent and replicable?

In her review and recommendations for ratings and rankings systems, Gisselquist suggests that such assessments need to be transparent and replicable because they have political consequences, in that assessments are used for political decision-making and sometimes allocation of resources and therefore require transparency to ensure all involved are aware of the inputs and methods used. The proposed framework, as it will be described in upcoming sections, will be transparent and replicable. Indeed, this book seeks to ensure the highest level of transparency and replicability of the assessment in order to guarantee its relevance, usefulness, and applicability to space actors while remaining inclusive and respectful of divergent interests and varying viewpoints on the exploration and use of outer space. This in particular requires that the framework accounts for different stakeholder groups, including civil society, government, and industry, and the most efficient way of doing this is by describing the framework in detail and offering it in a transparent manner for review in this book.

4.5 How sensitive and robust is the measure to different data and design choices?

Gisselquist (2014) notes "There are few hard rules about the 'best' data sources, indicators, methods of normalization, and weighting and aggregation. Well-designed indexes however describe and justify their choices in each of these areas and examine the impact of these choices on the robustness of results." Here, it is suggested that designing a measure is subjective in some manner; however, it remains incumbent on the developers of an assessment to explain their rationale as much as possible for the selection of indicators, source data, and weighting and aggregation of scores because "they [the index producers] have the burden of showing that they are not (at least intentionally) cherry picking," (Gisselquist 2014). The explanation and description of the framework contained in this book seeks to satisfy this requirement by explaining the rationale for the data and design choices, which again are based on internationally agreed and established law and norms found within UN-level treaty, resolution, and guideline documents. The explanation for the framework's design choices will be presented for each indicator in the upcoming sections.

4.6 Does the measure address key questions of interest?

This guideline is meant to ensure that the assessment actually measures what is purported to be measured, and that the assessment is used only in such a way that makes sense for the data and unit of analysis that is being assessed. "Index users should consider whether it captures empirically what is under investigation, including country coverage, time coverage, and the level of analysis at which measurement is taken…" (Gisselquist 2014). As explained previously, the framework makes clear the units of measure are at the country level, and initially only applicable to those countries with sufficiently available data to assess in terms of national space laws, regulations, and frameworks. Moreover, this framework emphasises its application on an annual basis with future work to feature its first-time application and, hopefully, subsequent annual use. Perhaps as the framework continues to be applied over time and the legitimacy of the rating grows, more countries will make their data available in the hope of receiving a good rating.

4.7 Does the measure fully capture governance in all its complexity?

The assessment proposed herein does not claim to capture governance of space activities in all its complexity, particularly given the resources needed for conducting the assessment remain constrained themselves to only existing and internationally established space law and norms. Given that space policy and space governance continues to evolve as more and more actors enter the domain, presently not all governance issues areas have been settled or, in some cases, even contemplated by the international community. Therefore, all aspects of governance are not captured, yet for those that are included in the assessment they will be fully detailed and addressed by the indicators in upcoming sections.

4.8 Does the measure behave as theory predicts?

This guideline is approached in two ways: assessing the measurement by comparing it against other measurements of the same concept, or assessing the measurement by seeing how it is connected to other concepts that are theoretically derived. Since the assessment proposed in this book is the first of its kind in the space policy literature, the first approach described here is moot; therefore, the second approach requires consideration in order to successfully answer this guideline question. To help place this guideline in the realm of ratings and rankings systems, Gisselquist (2014) offers the Word Bank's Worldwide Governance Indicators (WGI) as an example for "…exploring the relationship between governance and growth…" This is instructive for determining to which other concepts space policy is related and how compliance with international law relates to other theoretically derived relationships. In this way, the measurement would be useful for exploring the relationship between global governance and peace and security, which would suggest that, with increased compliance with global governance regimes, then peace and security in the outer space domain would improve.

For an answer to be found for this guideline question, the assessment framework would need to first be applied and the results thereof would require comparison with trends that endanger the fragile status quo of peace and security trends in outer space, such as the number of incidents of cooperation and the number of incidents of conflict or adversarial behaviour. Once this exercise is completed, only then would an answer would be available for determining whether the measure behaves as theory predicts. For example, if the assessment found most spacefaring countries were proving compliant with international space law and norms, then this should correspond with a peaceful and secure space domain. However, if there were a number of conflicts or adversarial actions taking place within the review period yet the measurement was demonstrating compliance on a large scale, the assessment would not be behaving as theory would predict since theory suggests compliance with international law and norms decreases conflict and adversarial behaviour. The assessment would require a baseline to demonstrate this change, thus the existing status quo of peace and security in outer space would serve as this.

4.9 How precisely can index values and confidence intervals be specified?

Whereas Gisselquist (2014) suggests that following this guideline is less important compared to others, she does note that "…index producers should somehow acknowledge the uncertainty and imprecision surrounding scores." In order to adhere to this guideline, the framework proposed here will adopt a similar approach as taken by the World Bank's WGI confidence intervals whereby "The estimation of the standard error is not based on the survey sample size, but rather on the number of assessments for each country and the degree to which their scores are consistent with each other," (Gisselquist 2014). While the compliance assessment will largely draw upon UN and country-specific source documents, other source data will be used, such as the *Space Security Index*, to complement and inform the ratings, and help to compare how different reports assess the same country.

4.10 Is the weighting correct?

The rationale for weighting in traditional governance ratings and rankings systems comes from two concepts, namely the degree of confidence in each component's accuracy and the relative importance of each component to governance. For purposes of this assessment, however, weighting will not be done in an effort to signal that certain components of space policy governance and adherence to international laws and norms are more important than other components. Rather, this assessment will weight its components based upon whether a law or norm is binding or non-binding, and in an effort to reflect an inclusive set of perspectives pertaining to outer space decision-making and policymaking. The assessment will weight in this manner as binding international law represents a more consequential, and thus more pressing, area for policymakers to follow, while also representing generally greater consensus and willingness to be held accountable at the international level as compared to non-binding norms established in UN resolutions, principles, and guidelines. Moreover, this book recognises that international law-making is often a process resulting from negotiations and agreements taking place among leading powers, excluding the perspectives of lesser powers or those involved on the periphery. According to Nico Krisch (2005), "Most predominant states have been active forces behind the development of international law, and they have made extensive use of the international legal order to stabilize and improve their position." Therefore, this assessment seeks to balance, through its weighting of different components measuring compliance, the importance of binding international law with the fact that the process that produces such law is often exclusionary and unrepresentative of broader interests within the international community.

5 FRAMEWORK FOR ASSESSING COMPLIANCE

Since this book has now explored the two movements from which our assessment framework emerges, namely the evidence-based programming and policymaking movement and the ratings and rankings movement, alongside the subject matter that will be assessed and the best practices for doing so, this chapter will now detail the assessment framework and its associated indicators, accompanied by a brief description of what is meant by each, together with their respective scoring and weighting schemes. Prior to embarking on this, it remains prudent to affirm the objective this assessment seeks to achieve, which is to understand countries' levels of compliance with international space law and norms, and to rate counties according to how well they comply. The purpose of this assessment is not only to inform policymakers, but also to introduce an accountability and advocacy tool useful for promoting behaviour that is in the spirit of the peaceful exploration and use of outer space by all countries, a well-established principle found throughout international treaties, resolutions, and guidelines regarding outer space.

5.1 General scoring of the framework

Figure 5.1 illustrates how the various scoring elements are compiled to arrive at an overall assessment score and category designation of highly compliant, mostly compliant, partially compliant, minimally compliant, and non-compliant for each country that is assessed.

Figure 5.1: Relationship of Assessment Framework Scoring Elements

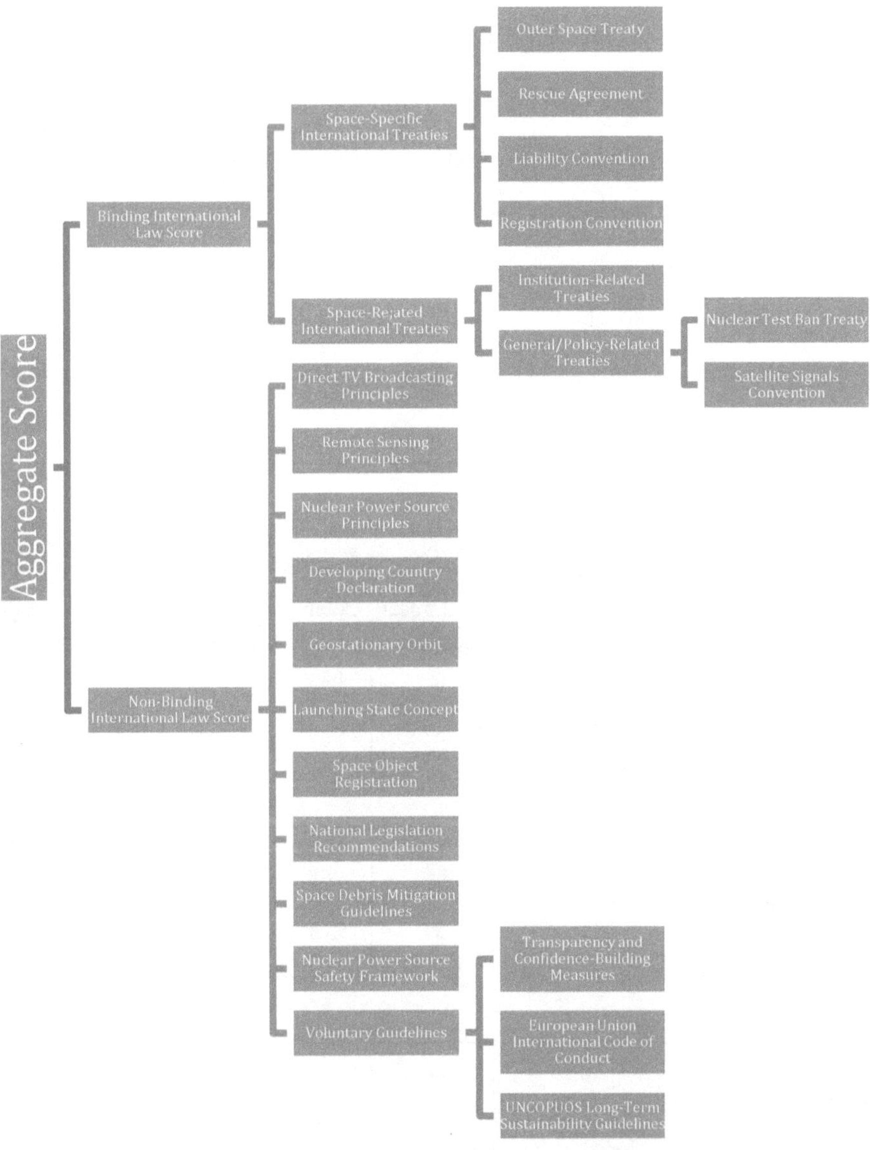

A comprehensive assessment framework is presented below in the form of a set of tables containing questions that assess levels of compliance among three different types of spacefaring countries: advanced, intermediate, and emerging.

Given the varying levels of space activities conducted by each of these different types of countries, some assessment areas may apply to some countries but not to others. As such, some questions are not applied to intermediate and/or emerging spacefaring countries, resulting in different scoring potentials for each tier (advanced, intermediate, or emerging). The ranges for the scoring potentials of each tier and the category intervals that determine whether countries are highly, mostly, partially, minimally, or non-compliant are presented below in Figure 5.2.

Figure 5.2: Scoring Ranges for Ratings						
	Interval	Highly Compliant	Mostly Compliant	Partially Compliant	Minimally Compliant	Non-Compliant
Advanced	8.11	40.55 – 32.45	32.44 – 24. 34	24.33 – 16.23	16.22 – 8.12	8.11 – 0.00
Intermediate	7.86	39.30 – 31.45	31.44 – 23. 59	23.58 – 15.73	15.72 – 7.87	7.86 – 0.00
Emerging	6.14	30.70 – 24.57	24.56 – 18.43	18.42 – 12.29	12.28 – 6.15	6.14 – 0.00

Figure 5.3: Legend for Applicable Questions Based on Space Activity

Applicable to Advanced spacefaring countries ONLY
Applicable to Advanced and Intermediate spacefaring countries ONLY
Applicable to Advanced, Intermediate, and Emerging spacefaring countries

A legend is provided in Figure 5.3 to guide how the assessment framework should be applied to which each of spacefaring country based upon its level of space activities. The questions presented in each table below are coloured in with the appropriate designation as to whether they apply only to advanced spacefaring countries, only to advanced and intermediate spacefaring countries, or to all advanced, intermediate, and emerging spacefaring countries.

Additionally, as evidenced below in the possible scoring for each question included in the framework, a range of possible replies is used to capture the various degrees and possible responses for each issue area, such as "always," "sometimes," or "never." For questions where "sometimes" is a possible response, the country should be scored the minimum amount of points possible (i.e. zero) in cases where the incident has only occurred once. "Sometimes" implies that the country has had at least more than one opportunity to demonstrate some level of compliance with international space law and norms. However, if the record only consists of one example and non-compliance was demonstrated, then the country should be assessed a score of zero for non-compliance.

For example, if only one opportunity existed for a country to provide assistance to an astronaut in distress and the country decided not to provide assistance, then a score of zero should be assessed. However, if two opportunities existed for a country to provide assistance to an astronaut in distress and it provided assistance in one case and refused to provide assistance in the second case, then a score of one for "sometimes" should be assessed.

Table A below describes the aggregate scoring potential for all questions related to the binding international law and non-binding international law categories within the assessment framework. The aggregate score is derived by summing the two category scores from Tables B.1 (binding international law score) and Table B.2 (non-binding international law score).

A	Aggregate Score
Binding International Law Score *(Enter total from last line of Table B.1)*	Advanced: XX / 20.15 Intermediate: XX / 18.90 Emerging: XX / 16.90
Non-Binding International Law Score *(Enter total from last line of Table B.2)*	Advanced: XX / 20.40 Intermediate: XX / 20.40 Emerging: XX / 13.80
Total *(Sum binding and non-binding international law scores)*	Advanced: XX / 40.50 Intermediate: XX / 39.30 Emerging: XX / 30.70

Table B.1 below captures the score for those sections relevant to binding international law instruments, including space-specific and space-related international treaties. The score for each section, space-specific international treaties and space-related international treaties, respectively, is weighted by 50 percent and then summed to arrive at the overall binding international law score.

B.1	Binding International Law Score
Space-specific International Treaties *(Use total from last line of Table B.1.1 and multiply by .50)*	Advanced: XX / 14.25 Intermediate: XX / 13.0 Emerging: XX / 11.0
Space-related International Treaties *(Use total from last line of Table B.1.2 and multiply by .50)*	Advanced: XX / 5.90 Intermediate: XX / 5.90 Emerging: XX / 5.90
Total *(Sum space-specific and space-related treaties scores and then bring total to Table A)*	Advanced: XX / 20.15 Intermediate: XX / 18.90 Emerging: XX / 16.90

Table B.1.1 below illustrates the scoring potential for all tiers of spacefaring countries (advanced, intermediate, and emerging) according to how well they comply with space-specific international treaties. For this section, the scores for each treaty that is assessed (i.e. the Outer

Space Treaty, the Rescue Agreement, the Liability Convention, and the Registration Convention) are weighted by 25 percent and then summed to arrive at the overall score for the space-specific international treaties. The colour cording in the table serves as a reminder that some assessment questions are only relevant to advanced and intermediate spacefaring countries, thus these questions are filled in according to the legend found in Figure 5.3 above.

B.1.1 Space-specific International Treaties	
(a) Treaty on Principles Governing the Activities of States in the Exploration and Use of Outer Space, including the Moon and Other Celestial Bodies *(25 percent of section score)*	
a. Has the country ratified or acceded to this treaty?	Yes, fully = 2 Yes, with reservations = 1 No = 0
b. Is the treaty domesticated in national laws or administrative regulations or procedures?	Full domestication = 2 Partial domestication = 1 No domestication = 0
c. Article I: Do country's laws or administrative regulations or procedures recognise the role of international law?	Full recognition = 2 Partial recognition = 1 No recognition = 0
d. Article I: Do the country's laws or administrative regulations or procedures affirm that the exploration and use of outer space should be carried out for the benefit and interests of all countries?	Full affirmation = 2 Partial affirmation = 1 No affirmation = 0
e. Article I: Do the country's laws or administrative regulations or procedures recognise outer space as the province of all mankind?	Full recognition = 2 Partial recognition = 1 No affirmation = 0
f. Article I: Do the country's laws or administrative regulations or procedures recognise that outer space should be free for exploration without discrimination and with free access to celestial bodies?	Full recognition = 2 Partial recognition = 1 No recognition = 0
g. Article I: Do the country's laws or administrative regulations or procedures recognise the freedom of scientific investigation in outer space?	Full recognition = 2 Partial recognition = 1 No recognition = 0
h. Article I: Does the country facilitate and participate in international scientific investigation?	Frequently = 2 Occasionally = 1 Never = 0
i. Article II: Do the country's laws or administrative regulations or procedures allow for, or encourage, the appropriation of outer space and celestial bodies by claims of sovereignty, use, or occupation?	Always = 0 Sometimes = 1 Never = 2
j. Article II: Does the country, by its actions or the actions of those within its jurisdiction, appropriate, or encourage the appropriation of, outer space and celestial bodies and resources by claims of sovereignty, use, or occupation?	Frequently = 0 Occasionally = 1 No = 2

Article III: Elements of this article are already captured in questions asked above, therefore no separate scoring is assessed.	
k. Article IV: Has the country placed into orbit or installed on celestial bodies weapons of mass destruction, including nuclear weapons?	Repeatedly = 0 Once = 1 No = 2
l. Article IV: Has the country established bases, installations, or fortifications, or tested weapons or conducted military manoeuvres on celestial bodies?	Repeatedly = 0 Once = 1 No = 2
m. Article IV: Do the country's laws or administrative regulations or procedures allow for the conduct of such types of activities as described in Question 'l' above?	Always = 0 Sometimes = 1 Never = 2
n. Article V: Has the country either provided assistance or rejected assistance to astronauts of other countries who are in distress?	Always provided / not applicable = 2 Sometimes provided = 1 Never provided = 0
o. Article V: Does the country proactively inform the United Nations and other countries regarding potentially dangerous phenomena to astronauts?	Always = 2 Sometimes = 1 Never = 0
p. Article VI: Does the country have in place an authorisation and supervision regime for space activities conducted by non-governmental entities?	Yes = 2 Partial = 1 No = 0
q. Article VII: Does the country recognise its liability as a launching state in its laws or administrative regulations or procedures?	Full recognition = 2 Partial recognition = 1 No recognition = 0
r. Article VII: Has the country denied its liability as a launching state for damage caused?	Always = 0 Sometimes = 1 Never = 2
s. Article VIII: Does the country maintain an updated national registry of launched space objects?	Always = 2 Sometimes = 1 Never = 0
t. Article VIII: Has the country ever returned a space object belonging to another country, but which landed in its jurisdiction?	Always / not applicable = 2 Sometimes = 1 Never = 0
u. Article IX: Do the country's laws or administrative regulations or procedures account for avoiding harmful contamination, including debris, of the space environment?	Full account = 2 Partial account = 1 No account = 0
v. Article IX: Do the country's laws or administrative regulations or procedures seek to limit damage to the Earth environment by extraterrestrial material?	Full account = 2 Partial account = 1 No account = 0
w. Article X: Does the country allow other countries to observe the flight of space objects that it launches?	Always = 2 Sometimes = 1 Never = 0
x. Article XI: Does the country inform the United Nations and other countries of its space activities, including the nature, conduct, location, and results of the activities?	Always = 2 Sometimes = 1 Never = 0

y. Article XII: Does the country allow access and visits to its outer space stations, installations, equipment, and vehicles, provided reasonable notification of a such visit is given by the visiting representatives?	Always / not applicable = 2 Sometimes = 1 Never = 0
Articles XIII, XIV, XV, XVI, and XVII: These are administrative articles, therefore no scoring is assessed.	
Treaty score *(Sum the above scores and multiply by .25)*	Advanced: XX / 12.50 Intermediate: XX / 10 Emerging: XX / 6.5
(b) Agreement on the Rescue of Astronauts, the Return of Astronauts and the Return of Objects Launched into Outer Space *(25 percent of section score)*	
a. Has the country ratified or acceded to this treaty?	Yes, in full = 2 Yes, with reservations = 1 No = 0
b. Is the treaty domesticated in national laws or administrative regulations or procedures?	Full domestication = 2 Partial domestication = 1 No domestication = 0
c. Article 1: Has the country notified the launching state(s) of astronauts who have suffered accidents, been in distress, or made emergency landings?	Always / not applicable = 2 Sometimes = 1 Never = 0
d. Article 1: Has the country notified the United Nations of astronauts who have suffered accidents, been in distress, or made emergency landings?	Always / not applicable = 2 Sometimes = 1 No = 0
e. Article 2: Has the country taken all possible steps to rescue and render all necessary assistance to astronauts who have suffered accidents, been in distress, or made emergency landings?	Always / not applicable = 2 Sometimes = 1 No = 0
f. Article 3: Has the country provided assistance in search and rescue operations for astronauts who have suffered accidents, been in distress, or made emergency landings?	Always / not applicable = 2 Sometimes = 1 No = 0
g. Article 4: Has the country safely and promptly returned to their launching states astronauts who have suffered accidents, been in distress, or made emergency landings?	Always / not applicable = 2 Sometimes = 1 No = 0
h. Article 5: Has the country notified the launching state and United Nations of a space object or part of a space object that has returned to Earth?	Always / not applicable = 2 Sometimes = 1 No = 0
i. Article 5: Has the country in which the returning space object landed practicably assisted to recover the object or its parts?	Always / not applicable = 2 Sometimes = 1 No = 0

j. Article 5: Has the country in which the returning space object landed returned the space object or its parts and provided their identifying data to the launching state?	Always / not applicable = 2 Sometimes = 1 No = 0
k. Article 5: Has the country in which the returning space object landed eliminated the possibility of its danger if it is found to be hazardous or deleterious?	Always / not applicable = 2 Sometimes = 1 No = 0
l. Article 5: Has the launching state paid for expenses related to the recovery and return of space object or its part incurred by the country in which the space object landed?	Always / not applicable = 2 Sometimes = 1 No = 0
Articles 6, 7, 8, 9, and 10: These are administrative articles, therefore no scoring is assessed.	
Treaty score *(Sum the above scores and multiply by .25)*	Advanced: XX / 6 Intermediate: XX / 6 Emerging: XX / 6
(c) Convention on International Liability for Damage Caused by Space Objects *(25 percent of section score)*	
a. Has the country ratified or acceded to this treaty?	Yes, in full = 2 Yes, with reservations = 1 No = 0
b. Is the treaty domesticated in national laws or administrative regulations or procedures?	Full domestication = 2 Partial domestication = 1 No domestication = 0
Article I provides for definitions of terms used in the treaty, therefore no scoring is assessed.	
c. Article II: Do the country's laws or administrative regulations or procedures recognise its liability to pay compensation for damage caused by its space objects on the Earth's surface or aircraft in flight?	Full recognition = 2 Partial recognition = 1 No recognition = 0
d. Article II: Has the country refused to pay compensation in practice for damage caused by its space objects on the Earth's surface or aircraft in flight?	Always = 0 Sometimes = 1 Never / not applicable = 2
e. Article III: Do the country's laws or administrative regulations or procedures recognise its liability owing to its fault or fault of persons for which it is responsible in the case of damage to another space object, persons, or property in space of another launching state?	Full recognition = 2 Partial recognition = 1 No recognition = 0
f. Article III: Has the country refused to accept liability in the case of damage to another space object, persons, or property in space of another launching state caused by its fault or fault of persons for which it is responsible?	Always = 0 Sometimes = 1 Never / not applicable = 2
g. Article IV: Do the country's laws or administrative regulations or procedures recognise its absolute joint and several liability with a second-party launching state if its space object caused damage to the second-party launching state's	Full recognition = 2 Partial recognition = 1 No recognition = 0

space object, which then resulted in damage on the Earth's surface or aircraft in flight of a third state?	
h. Article IV: Does the country's laws or administrative regulations or procedures recognise its joint and several liability apportioned according to fault with a second-party launching state if its space object caused damage to the second-party launching state's space object, which then resulted in damage to a third state's space object, person, or property in space?	Full recognition = 2 Partial recognition = 1 No recognition = 0
i. Article IV: Has the country refused to pay compensation according to its fault in cases of joint and several liability for damage caused to a third-party state?	Always = 0 Sometimes = 1 Never / not applicable = 2
j. Article V: Do the country's laws or administrative regulations or procedures recognise joint and several liability in cases when it and another state jointly launch a space object?	Full recognition = 2 Partial recognition = 1 No recognition = 0
k. Article V: Do country's laws or administrative regulations or procedures account for its roles a joint launch participant if a space object is launched from its territory or facility?	Full account = 2 Partial account = 1 No account = 0
Articles VI and VII sets forth conditions under which other articles apply and is not actionable by states, therefore no scoring is assessed.	
l. Article VIII: Do the country's laws or administrative regulations or procedures provide for a claims and compensation process for damage which it may suffer by other launching states?	Full provision = 2 Partial provision = 1 No provision = 0
Article IX, X, XI, XII, and XIII prescribe the administrative details through which claims and compensation should be processed, therefore no scoring is assessed.	
Articles XIV, XV, XVI, XVII, and XVIII prescribe the administrative details through which a Claims Commission should be established and make adjudications in the case of no settlements through diplomatic channels established elsewhere in the treaty, therefore no scoring is assessed.	
m. Article XIX: Has the country refused or failed to pay compensation or an award decided by a Claims Commission as set forth in other treaty articles?	Always = 0 Sometimes = 1 Never / not applicable = 2
Article XX prescribes the process through which the expenses associated with the Claims Commission should be borne by state participants, therefore no scoring is assessed.	
n. Article XXI: Has the country rendered appropriate and rapid assistance in cases when a space object presents large-scale danger to human life or interferes with a population's living conditions or functioning of a vital centre?	Always / not applicable = 2 Sometimes = 1 Never = 0

Articles XXII, XXIII, XXIV, XXV, XXVI, XXVII, XXVIII: These are administrative articles, therefore no scoring is assessed.	
Treaty score *(Sum the above scores and multiply by .25)*	Advanced: XX / 7 Intermediate: XX / 7 Emerging: XX / 7
(d) Convention on Registration of Objects Launched into Outer Space *(25 percent of section score)*	
a. Has the country ratified or acceded to this treaty?	Yes, in full = 2 Yes, with reservations = 1 No = 0
b. Is the treaty domesticated in national laws or administrative regulations or procedures?	Full domestication = 2 Partial domestication = 1 No domestication = 0
Article I provides for definitions of terms used in the treaty, therefore no scoring is assessed.	
c. Article II: Does the country maintain an updated national registry of launched space objects that it launches and that it launches jointly with other launching states?	Always / not applicable = 2 Sometimes = 1 Never = 0
Article III describes actions to be taken by the United Nations rather than a specific country, therefore no scoring is assessed.	
d. Article IV: Does the country provide detailed information to the United Nations for its launched space objects, including name of the country, space object registration number, date and territory of launch, basic orbital parameters, and general function of the space object?	Full information / not applicable= 2 Partial information = 1 No information = 0
e. Article IV: Does the country update the United Nations on space objects for which it previously shared information, but which are no longer in Earth orbit?	Always / not applicable = 2 Sometimes = 1 Never = 0
Article V describes a procedural administrative action, therefore no scoring is assessed.	
f. Article VI: Does the country assist with providing identifying information it has gained through space monitoring and tracking facilities for space objects that are unidentifiable and which are of a hazardous or deleterious nature?	Always = 2 Sometimes = 1 Never = 0
Articles VII, VIII, IX, X, and XII: These are administrative articles, therefore no scoring is assessed.	
Treaty score *(Sum the above scores and multiply by .25)*	Advanced: XX / 3 Intermediate: XX / 3 Emerging: XX / 3
Total *(Sum the above treaty scores and bring total to Table B.1)*	Advanced: XX / 28.50 Intermediate: XX / 26.0 Emerging: XX / 22.0

Table B.1.2 below illustrates the scoring potential for space-related international treaties. The Nuclear Weapon Test Ban Treaty and the Satellite Signals Convention are considered for this section, as well as institution-related treaties. The Nuclear Test Ban Treaty and the Satellite Signals Convention are accorded a 70 percent weighting for this section, while the score resulting from the institution-related treaties are weighted 30 percent. Once weighted, the scores are summed to produce the overall score for the space-related international treaties section, which is used in Table B.1 above.

B.1.2 Space-related International Treaties	
(a) Treaty Banning Nuclear Weapon Tests in the Atmosphere, in Outer Space and under Water	
a. Has the country ratified or acceded to this treaty?	Yes, fully = 2 Yes, with reservations = 1 No = 0
b. Is the treaty domesticated in national laws or administrative regulations or procedures?	Full domestication = 2 Partial domestication = 1 No domestication = 0
iii. Article I: Do the country's laws or administrative regulations or procedures prohibit, prevent, and forbid the carrying out of nuclear weapon tests at places under its jurisdiction or control, including in the atmosphere and outer space and underwater?	Full prohibition = 2 Partial prohibition = 1 No prohibition = 0
iv. Article I: Since implementing the treaty, has the country carried out a nuclear weapon test at a place under its jurisdiction or control, including in the atmosphere and outer space and underwater?	Repeatedly = 0 Once = 1 Never = 2
Articles II, III, IV, and V: These are administrative articles, therefore no scoring is assessed.	
Treaty score *(Sum the above scores)*	Advanced: XX / 8 Intermediate: XX / 8 Emerging: XX / 8
(b) Convention Relating to the Distribution of Programme-Carrying Signals Transmitted by Satellite	
a. Has the country ratified or acceded to this treaty?	Yes, in full = 2 Yes, with reservations = 1 No = 0
b. Is the treaty domesticated in national laws or administrative regulations or procedures?	Full domestication = 2 Partial domestication = 1 No domestication = 0

Article 1 provides for definitions of terms used in the treaty, therefore no scoring is assessed.	
iii. Article II: Do the country's laws or administrative regulations or procedures prevent the distribution on or from its territory programme-carrying signals by distributors for whom the emitted signal to or passing through the satellite is not intended?	Always = 2 Sometimes = 1 Never = 0
iv. Article II: Has the country, by its actions or the actions of those within its jurisdiction, allowed for the distribution on or from its territory programme-carrying signals by distributors for whom the emitted signal to or passing through the satellite is not intended?	Always = 0 Sometimes = 1 Never = 2
Articles 3, 4, 5, 6, and 7 describe the conditions in which Article 2 applies, therefore no scoring is assessed.	
Articles 8, 9, 10, 11 and 12: These are administrative articles, therefore no scoring is assessed.	
Treaty score *(Sum the above scores)*	Advanced: XX / 8 Intermediate: XX / 8 Emerging: XX / 8
(c) Institution-related treaties	
Individual treaties within this subsection are not scored as they are institution-specific and in some case outdated and therefore irrelevant. However, collectively if a country ratified or acceded to at least *two* of these treaties, it will be assessed a score of 2 to demonstrate its participation and cooperation in international legal instruments and organisations related to space, a key principle promoted in international space law documents. Therefore, the following treaties for consideration are:	
i. Agreement Relating to the International Telecommunications Satellite Organization (ITSO)	
ii. Agreement on the Establishment of the INTERSPUTNIK International System and Organization of Space Communications	
iii. Convention for the Establishment of a European Space Agency (ESA)	
iv. Agreement of the Arab Corporation for Space Communications (ARABSAT)	
v. Agreement on Cooperation in the Exploration and Use of Outer Space for Peaceful Purposes (INTERCOSMOS)	
vi. Convention on the International Mobile Satellite Organization	
vii. Convention Establishing the European Telecommunications Satellite Organization (EUTELSAT)	
viii. Convention for the Establishment of a European Organization for the Exploitation of Meteorological Satellites (EUMETSAT)	
ix. International Telecommunication Constitution and Convention	
Treaty score *(Multiply score by .30)*	Advanced: XX / 0.6 Intermediate: XX /0.6 Emerging: XX / 0.6

Total	Advanced: XX /11.80
(Sum the treaty scores from (a) and (b) and multiply by .70 and then sum with (c) score; bring total to Table B.1)	Intermediate: XX / 11.80 Advanced: XX / 11.80

Table B.2 below captures the score for those elements comprising non-binding international law, including the various principles, resolutions, and guidelines developed by the United Nations and other international organisations. The score achieved in Table B.2 is used in Table A to arrive at the aggregate score. Similar to the binding international law questions, the legend presented in Figure 5.3 is used to determine which questions apply to which tier of country (i.e. advanced, intermediate, or emerging) in the questions posed in Tables B.2.1 – B.2.10.

The scores from Tables B.2.1 – B.2.10 are each weighted by 10 percent in Table B.2, whereas the score produced from Table B.2.11 on voluntary guidelines remains unweighted in Table B.2. The different treatment of the tables is done intentionally to accord the appropriate emphasis to UN-level principles, resolutions, and guidelines and the comprehensive voluntary guidelines, which themselves are a result of significant international consensus-making for supporting the future development of space policy and space legal frameworks.

B.2 Non-Binding International Law Score	
(a) Principles Governing the Use by States of Artificial Earth Satellites for International Direct Television Broadcasting *(Enter total from Table B.2.1 and multiply by .10)*	Advanced: XX / 1.80 Intermediate: XX / 1.80 Emerging: XX / 1.80
(b) Principles Relating to Remote Sensing of the Earth from Outer Space *(Enter total from Table B.2.2 and multiply by .10)*	Advanced: XX / 1.80 Intermediate: XX / 1.80 Emerging: XX / 1.80
(c) Principles Relevant to the Use of Nuclear Power Sources in Outer Space *(Enter total from Table B.2.3 and multiply by .10)*	Advanced: XX / 3.0 Intermediate: XX / 3.0 Emerging: XX / 0.0
(d) Declaration on International Cooperation in the Exploration and Use of Outer Space for the Benefit and in the Interest of All States, Taking into Particular Account the Needs of Developing Countries *(Enter total from Table B.2.4 and multiply by .10)*	Advanced: XX / 1.0 Intermediate: XX / 1.0 Emerging: XX / 0.0
(e) International cooperation in the peaceful uses of outer space: Some aspects concerning the use of the geostationary orbit *(Enter total from Table B.2.5 and multiply by .10)*	Advanced: XX / 0.4 Intermediate: XX / 0.4 Emerging: XX /0.4

(f) Application of the concept of the "launching State" *(Enter total from Table B.2.6 and multiply by .10)*	Advanced: XX / 0.6 Intermediate: XX / 0.6 Emerging: XX / 0.6
(g) Recommendations on enhancing the practice of States and international intergovernmental organizations in registering space objects *(Enter total from Table B.2.7 and multiply by .10)*	Advanced: XX / 0.8 Intermediate: XX / 0.8 Emerging: XX / 0.4
(h) Recommendations on national legislation relevant to the peaceful exploration and use of outer space *(Enter total from Table B.2.8 and multiply by .10)*	Advanced: XX / 1.40 Intermediate: XX / 1.40 Emerging: XX / 1.20
(i) Space Debris Mitigation Guidelines of the Committee on the Peaceful Uses of Outer Space *(Enter total from Table B.2.9 and multiply by .10)*	Advanced: XX / 1.40 Intermediate: XX / 1.40 Emerging: XX / 1.40
(j) Safety Framework for Nuclear Power Source Applications in Outer Space *(Enter total from Table B.2.10 and multiply by .10)*	Advanced: XX / 2.0 Intermediate: XX / 2.0 Emerging: XX / 0.0
(k) Voluntary guidelines for transparency and confidence-building, space code of conduct, and long-term sustainability of space *(Enter total from Table B.2.11)*	Advanced: XX / 6 Intermediate: XX / 6 Emerging: XX / 6
Total *(Sum scores from above and bring total to Table A)*	Advanced: XX / 20.4 Intermediate: XX / 20.4 Emerging: XX / 13.8

B.2.1	**Principles Governing the Use by States of Artificial Earth Satellites for International Direct Television Broadcasting**
1. Are the country's satellite direct television broadcasting activities respectful of state sovereignty, and do they follow the principle of non-intervention?	Always / not applicable = 2 Sometimes = 1 Never = 0
2. Do the country's satellite direct television broadcasting activities promote the free dissemination and exchange of information and knowledge in cultural and scientific fields?	Always / not applicable = 2 Sometimes = 1 Never = 0
3. Does the country's satellite direct television broadcasting activities assist in education and social and economic development?	Always / not applicable = 2 Sometimes = 1 Never = 0
4. Are the country's satellite direct television broadcasting activities carried out on a friendly and cooperative basis with other countries?	Always / not applicable = 2 Sometimes = 1 Never = 0
5. Does the country have in a place an authorisation system through its laws or administrative regulations or procedures for the conduct of satellite direct television activities?	Full implementation = 2 Partial implementation = 1 No implementation = 0

6. Does the country consult with other states regarding broadcasting or receiving of satellite direct television broadcasting for purposes of coordinating the same service?	Always / not applicable = 2 Sometimes = 1 Never = 0
7. Does the country participate in agreements with other states for the protection of copyright and similar rights?	Always / not applicable = 2 Sometimes = 1 Never = 0
8. Does the country notify the United Nations of satellite direct television broadcasting?	Always / not applicable = 2 Sometimes = 1 Never = 0
9. Does the country notify and consult the receiving state of a satellite direct television signal of its intent to broadcast?	Always / not applicable = 2 Sometimes = 1 Never = 0
Principles score *(Sum scores from above and bring to Table B.2)*	Advanced: XX / 8 Intermediate: XX/ 8 Emerging: XX / 8

B.2.2	Principles Relating to Remote Sensing of the Earth from Outer Space
1. Does the country carry out remote sensing activities for the benefit and in the interests of all countries?	Always = 2 Sometimes = 1 Never = 0
2. Does the country's remote sensing activities observe the principle of full and permanent sovereignty of all states over their own wealth and natural resources?	Always = 2 Sometimes = 1 Never = 0
3. Does the country allow for international cooperation and participation in its remote sensing activities?	Always = 2 Sometimes = 1 Never = 0
4. Does the country enter into agreements with other states for the establishment and operation of data collection and storage stations and processing and interpretation facilities?	Always = 2 Sometimes = 1 Never = 0
5. Does the country offer technical assistance on a regular basis to other states interested in conducting remote sensing activities?	Always = 2 Sometimes = 1 No = 0
6. Does the country inform the United Nations of its remote sensing activities?	Always = 2 Sometimes = 1 Never = 0
7. Does the country disclose information to other states derived from remote sensing that could help avert harm to the Earth's natural environment?	Always = 2 Sometimes = 1 Never = 0

8. Does the country disclose information to other states derived from processing and analysing remote sensing data that is useful to states affected, or likely to be affected, by natural disasters?	Always = 2 Sometimes = 1 Never = 0
9. Does the country make available to remotely sensed states primary and processed data on the sensed state according to non-discriminatory and reasonable cost terms?	Always = 2 Sometimes = 1 Never = 0
10. Does the country allow for opportunities for a remotely sensed state to participate and benefit from activities that remotely sense it, including but not limited to access to remotely sensed data and participation in remote sensing activities ?	Always = 2 Sometimes = 1 Never = 0
Principles score *(Sum scores from above and bring to Table B.2)*	Advanced: XX / 18 Intermediate: XX / 18 Emerging: XX / 18

B.2.3	Principles Relevant to the Use of Nuclear Power Sources in Outer Space
1. Does the country design and use space objects with nuclear power sources with high degrees of confidence to avoid hazardous levels associated with operational or accidental circumstances?	Always / not applicable = 2 Sometimes = 1 Never = 0
2. Do the country's space objects with nuclear power sources follow design and construction that take into account generally accepted international radiological protection guidelines?	Always / not applicable = 2 Sometimes = 1 Never = 0
3. Do the country's space objects with nuclear power sources possess the capability of correcting or counteracting failures or malfunctions?	Always / not applicable = 2 Sometimes = 1 Never = 0
4. Do the country's space objects with nuclear power sources include systems that follow practices of redundancy, physical separation, functional isolation, and adequate independence of components?	Always / not applicable = 2 Sometimes = 1 Never = 0
5. Does the country only operate nuclear power sources on space objects that are on interplanetary missions, in sufficiently high orbits, and in low-Earth orbits that are later stored in sufficiently high orbits after their mission operations?	Always / not applicable= 2 Sometimes = 1 Never = 0
6. Does the country's space objects with nuclear power sources only use highly enriched uranium 235 as fuel?	Always / not applicable = 2 Sometimes = 1 Never = 0

7. Does the country only make critical its nuclear reactors once they have reached their operating orbit or interplanetary trajectory?	Always / not applicable= 2 Sometimes = 1 Never = 0
8. Does the country design and constructs its satellites with nuclear power sources to ensure that they do not become critical before reaching their operating orbit or interplanetary trajectory?	Always / not applicable = 2 Sometimes = 1 Never = 0
9. Does the country's space objects with nuclear reactors include an effective and controlled disposal of the nuclear reactor for those in an operational orbit less than sufficiently high?	Always / not applicable = 2 Sometimes = 1 Never = 0
10. Does the country only use radioisotope generators in space objects that are in interplanetary missions, or in missions in Earth orbit but which are later disposed of in high orbit?	Always / not applicable = 2 Sometimes = 1 Never = 0
11. Does the country's space object with radioisotope generators include containment systems that can withstand the heat and aerodynamic forces of re-entry, and which prevent the scattering of radioactive material in the environment?	Always / not applicable = 2 Sometimes = 1 Never = 0
12. Does the country conduct a comprehensive safety assessment for all phases of a mission for a space object with a nuclear power source?	Always / not applicable = 2 Sometimes = 1 Never = 0
13. Does the country indicate to the public and inform the United Nations of when it will launch a space object with a nuclear power source?	Always / not applicable = 2 Sometimes = 1 Never = 0
14. Does the country provide timely and updated information on space objects with nuclear power sources that malfunction and which may re-enter?	Always / not applicable = 2 Sometimes = 1 Never = 0
15. Does the country provide assistance to eliminate actual and possible harmful effects of re-entered space objects with nuclear power sources?	Always / not applicable = 2 Sometimes = 1 Never = 0
Principles score *(Sum scores from above and bring to Table B.2)*	Advanced: XX / 30 Intermediate: XX / 30 Emerging: XX / 0

B.2.4	Declaration on International Cooperation in the Exploration and Use of Outer Space for the Benefit and in the Interest of All States, Taking into Particular Account the Needs of Developing Countries	
1. Does the country engage in equitable and mutually acceptable cooperation with other states according to fair and reasonable contractual terms?	Always = 2 Sometimes = 1 Never = 0	
2. Does the country offer developing countries and countries with incipient space programmes opportunities to participate in advanced space activities?	Always = 2 Sometimes = 1 Never = 0	
3. Do the country's space activities promote the development of space science and technology and its applications?	Always = 2 Sometimes = 1 Never = 0	
4. Do the country's space activities help develop relevant and appropriate capabilities in states interested in space?	Always = 2 Sometimes = 1 Never = 0	
5. Do the country's space activities facilitate the exchange of expertise and technology on a mutually acceptable basis?	Always = 2 Sometimes = 1 No = 0	
Declaration score *(Sum scores from above and bring to Table B.2)*	Advanced: XX / 10 Intermediate: XX / 10 Emerging: XX / 0	

B.2.5	International cooperation in the peaceful uses of outer space: Some aspects concerning the use of the geostationary orbit	
1. Does the country already having access to a geostationary orbit/spectrum take all practicable steps to enable a developing country, which wishes to use the same, to have equitable access to the requested orbit/spectrum?	Always / not applicable = 2 Sometimes = 1 Never = 0	
2. Does the country file satellite orbit and frequency requests according to International Telecommunications Union regulations?	Always / not applicable = 2 Sometimes = 1 Never = 0	
Resolution score *(Sum scores from above and bring to Table B.2)*	Advanced: XX / 4 Intermediate: XX / 4 Emerging: XX / 4	

B.2.6	Application of the concept of the "launching State"	
1. Does the country have in place national laws authorizing and providing for continuing supervision of non-governmental entities operating in outer space?	Full implementation = 2 Partial implementation = 1 No implementation = 0	
2. Does the country voluntarily provide information for the on-orbit transfer of ownership of space objects?	Always = 2 Sometimes = 1 Never = 0	

3. Does the country have harmonized national space laws with international space laws?	Full harmonization = 2 Partial harmonization = 1 No harmonization = 0
Resolution score *(Sum scores from above and bring to Table B.2)*	Advanced: XX / 6 Intermediate: XX / 6 Emerging: XX / 6

B.2.7	Recommendations on enhancing the practice of States and international intergovernmental organizations in registering space objects
1. Does the country provide detailed information on its registered space objects to the United Nations, including: international designators, Coordinated Universal Time of launch date, kilometres and minutes and degrees of orbital parameters, geostationary orbit location (as applicable), change in status of operations, date decay or re-entry, date and condition of moving to a disposal orbit, and web links to space object information?	All information = 2 Partial information = 1 No information = 0
2. Does the country from which the territory or facility a space object is launched jointly determine with other states or intergovernmental organisations which entity should register the space object?	Always / not applicable = 2 Sometimes = 1 Never = 0
3. If participating in a joint launch, does the country register the space object separately from its joint launching state?	Always / not applicable = 2 Sometimes = 1 Never = 0
4. If a space object changes supervision, does the country of registry provide to the United Nations detailed information, including date of change in supervision, identification of the new owner or operator, change in orbital position, and change of function?	All information = 2 Partial information = 1 No information = 0
Recommendations score *(Sum scores from above and bring to Table B.2)*	Advanced: XX / 8 Intermediate: XX / 8 Emerging: XX / 4

B.2.8	Recommendations on national legislation relevant to the peaceful exploration and use of outer space
1. Does the country have a national framework that governs the launch of objects into and their return from outer space, the operation of launch or re-entry sites and the operation and control of space objects in orbit, design and manufacture of spacecraft, application of space science and technology, and exploration activities and research?	All = 2 Some = 1 None = 0
2. Does the country issue authorisations and ensure supervision over space activities carried out from its jurisdiction or elsewhere when carried out by its citizens or legally established persons?	Always = 2 Sometimes = 1 Never = 0

3. Does the country have a national authority that authorizes space activities, including procedures for granting, modifying, suspending, and revoking authorisations?	All procedures = 2 Some procedures = 1 No procedures = 0
4. Does the country's authorisation regime ensure the safe conduct and minimal risk to persons for space activities, including assessing the expertise and technical qualifications of space operators and requiring adherence to safety and technical standards in line with the Space Debris Mitigation Guidelines?	Fully ensures = 2 Partially ensures = 1 Does not ensure = 0
5. Does the country's supervision and authorisation regime include on-site inspections or general reporting requirements?	Fully includes = 2 Partially includes = 1 Does not include = 0
6. Does the country's laws or administrative regulations or procedures offer recourse from operators or owners of space objects if the country's liability is engaged in the case of damages, such as through the use of insurance requirements or indemnification procedures?	Full recourse = 2 Partial recourse = 1 No recourse = 0
7. Does the country's supervision regime apply in case of transfer of ownership of space objects of non-governmental entities?	Always / not applicable = 2 Sometimes = 1 Never = 0
Recommendations score *(Sum scores from above and bring to Table B.2)*	Advanced: XX / 14 Intermediate: XX / 14 Emerging: XX / 12

B.2.9	Space Debris Mitigation Guidelines of the Committee on the Peaceful Uses of Outer Space
1. Do the country's laws or administrative regulations or procedures require space systems to be designed to not release debris during normal operations?	Always = 2 Sometimes = 1 Never = 0
2. Do the country's laws or administrative regulations or procedures require spacecraft and launch vehicle orbital stages to be designed to avoid accidental break-ups?	Always = 2 Sometimes = 1 Never = 0
3. Do the country's laws or administrative regulations or procedures require spacecraft and launch vehicle orbital stages to have disposal and passivation measures in the case of failure?	Always = 2 Sometimes = 1 Never = 0
4. Do the country's laws or administrative regulations or procedures require spacecraft operators to follow collision avoidance procedures through adjustment of launch times and on-orbit avoidance manoeuvres?	Always = 2 Sometimes = 1 Never = 0
5. Does the country's laws or administrative regulations or procedures require space objects to deplete or make safe their on-board sources of stored energy, including through passivation by removal of residual propellants, compressed fluids, and electrical storage devices?	Always = 2 Sometimes = 1 Never = 0

6. Do the country's laws or administrative regulations or procedures limit the long-term presence of space objects in low-Earth orbit after their end of mission, including through controlled removal from orbit through re-entry or in orbits beyond low-Earth orbit?	Always = 2 Sometimes = 1 Never = 0
7. Do the country's laws or administrative regulations or procedures limit the long-term presence of space objects in geosynchronous Earth orbit after their end of mission, including through controlled removal to orbits beyond geosynchronous Earth orbit?	Always = 2 Sometimes = 1 Never = 0
Guidelines score *(Sum scores from above and bring to Table B.2)*	Advanced: XX / 14 Intermediate: XX / 14 Emerging: XX / 14

B.2.10	Safety Framework for Nuclear Power Source Applications in Outer Space
1. Does the country have in place safety policies, requirements, and processes that protect people and the environment in Earth's biosphere from potential hazards associated with relevant launch, operation, and end-of-service phases of space nuclear power source applications?	All phases / not applicable = 2 Some phases = 1 No phases = 0
2. Does the country's space activities authorisation regime verify the rationale and require justification for the use of space nuclear power source applications?	Always / not applicable = 2 Sometimes = 1 Never = 0
3. Does the country's space launch authorisation regime require an independent safety evaluation assessing the risk to people and the environment for launches, operations, and end-of-service phases for space nuclear power source applications?	Always / not applicable = 2 Sometimes = 1 Never = 0
4. Does the country conduct emergency preparedness activities, including emergency planning, training, rehearsals and development of procedures and communication protocols, in preparation for radiation exposure to people and the environment as a result of space nuclear power applications?	Repeatedly = 2 Once = 1 Never = 0
5. Does the country's laws or administrative regulations or procedures identify space operators of space nuclear power source applications as the primary responsibility holders for operations of such applications?	Full recognition = 2 Partial recognition = 1 No recognition = 0
6. Does the country's laws or administrative regulations or procedures require space operators to include safety management to form part of overall space mission management?	Always / not applicable = 2 Sometimes = 1 Never = 0

7. Does the country's laws or administrative regulations or procedures require space operations to have technical competence in nuclear safety, including qualified individuals and facilities for designing, testing, and analysing nuclear safety capabilities part of space missions?	Always / not applicable = 2 Sometimes = 1 Never = 0
8. Does the country's laws or administrative regulations or procedures require space operators to integrate safety considerations from design to development to launch and operations and end-of-service for the entire space nuclear power source application?	Always / not applicable = 2 Sometimes = 1 Never = 0
9. Does the country's laws or administrative regulations or procedures require space operators to conduct risk assessments on the launch, operation, and end-of-service phases of space nuclear power source applications?	Always / not applicable = 2 Sometimes = 1 Never = 0
10. Does the country's laws or administrative regulations or procedures require mitigation measures for accidents of space nuclear power source applications with the potential to release radioactive material into the environment?	Always / not applicable = 2 Sometimes = 1 Never = 0
Framework score *(Sum scores from above and bring to Table B.2)*	Advanced: XX / 20 Intermediate: XX / 20 Emerging: XX / 0

B.2.11	**Voluntary guidelines for transparency and confidence-building, space code of conduct, and long-term sustainability of space**
Individual recommendations included within these three separate sets of guidelines are not scored. However, if a country has implemented a majority of the individual recommendations contained within each set of guidelines in either its legal or administrative practice, it will be assessed a score of 2 per set of guidelines. Therefore, the following sets of guidelines for consideration are:	
i. Recommendations from the Group of Governmental Experts on Transparency and Confidence-Building Measures	
ii. Recommendations from the European Union's International Code of Conduct on Outer Space Activities	
iii. Recommendations from the UNCOPUOS Long-Term Sustainability Guidelines	
Guidelines score *(Sum scores from above and bring to Table B.2)*	Advanced: XX / 6 Intermediate: XX / 6 Emerging: XX / 6

5.2 Application of the framework to three test cases

Given that the above tables present the general scoring of the framework, it remains necessary to apply the framework in practice to demonstrate its utility. As suggested previously, there are three tiers of spacefaring countries to which the framework will be applied, namely advanced, intermediate, and emerging countries. Based on the results from each assessment within each tier, a country can be assigned a rating of highly compliant, mostly compliant, partially compliant, minimally compliant, or non-compliant. As a reminder, the scoring ranges for each rating category according to whether a country is advanced, intermediate, or emerging is presented above in Figure 5.2.

One example from each tier has been selected to apply the framework and determine the countries' levels of compliance with international space law and norms. These are the United States (advanced), South Korea (intermediate), and South Africa (emerging). Each example represents varying levels of space activities and unique developmental and regional differences.

5.2.1 Scoring of the United States

The United States received an overall score of 33.33 out of a total possible score of 40.55. This suggests the United States is a highly compliant country in terms of international space law and norms as it scores in the highest tier of possible scores, although it risks falling to the next rating below of mostly compliant given how closely its currently sits to the lower limit of the highly compliant rating. Specifically, it received a score of 19.03 out a of possible score of 20.15 for the binding international law category, and it received a score of 14.30 out of a possible score of 20.40 for the non-binding law category. Generally, the scores suggest the

United States is a good actor in space and follows accepted international space law and norms. However, the country performs better in terms of its formal, legally binding obligations found within international treaties as compared to following non-binding norms contained within UN principles and resolutions and guidelines. If it were to be "graded," the United States would receive a 94 percent grade on the binding international law category and a 70 percent grade on the non-binding category. This demonstrates that the United States's performance on the binding international law category helped carry its overall score in to the highly compliant scoring range. In particular, the United States misses points in the non-binding category on the Principles Governing the Use by States of Artificial Earth Satellites for International Direct Television Broadcasting and the Principles Relating to Remote Sensing of the Earth from Outer Space. This is largely due to the United States conducting satellite broadcasting with governmental information to geopolitical adversaries. Similarly, the United States holds a well-documented reputation for conducting remote sensing for its intelligence and surveillance operations (i.e. not for the benefit and in the interests of geopolitical adversaries who are remotely sensed), an area where most, if not all, states that possess remote sensing capabilities likely falter in adhering to the remote sensing principles. If the United States wishes to maintain its highly compliant rating and to improve its overall score, the framework suggests that improvements could be made in adopting and abiding by additional non-binding guidelines and norms.

Figure 5.4 below presents the scoring of the United States for the various elements comprising the assessment framework, showing the assessed score out of the potential score possible for each element. Appendix I includes the detailed scoring for each question within the framework applied to the United States.

Figure 5.4: Scoring of Assessment Framework Elements for the United States

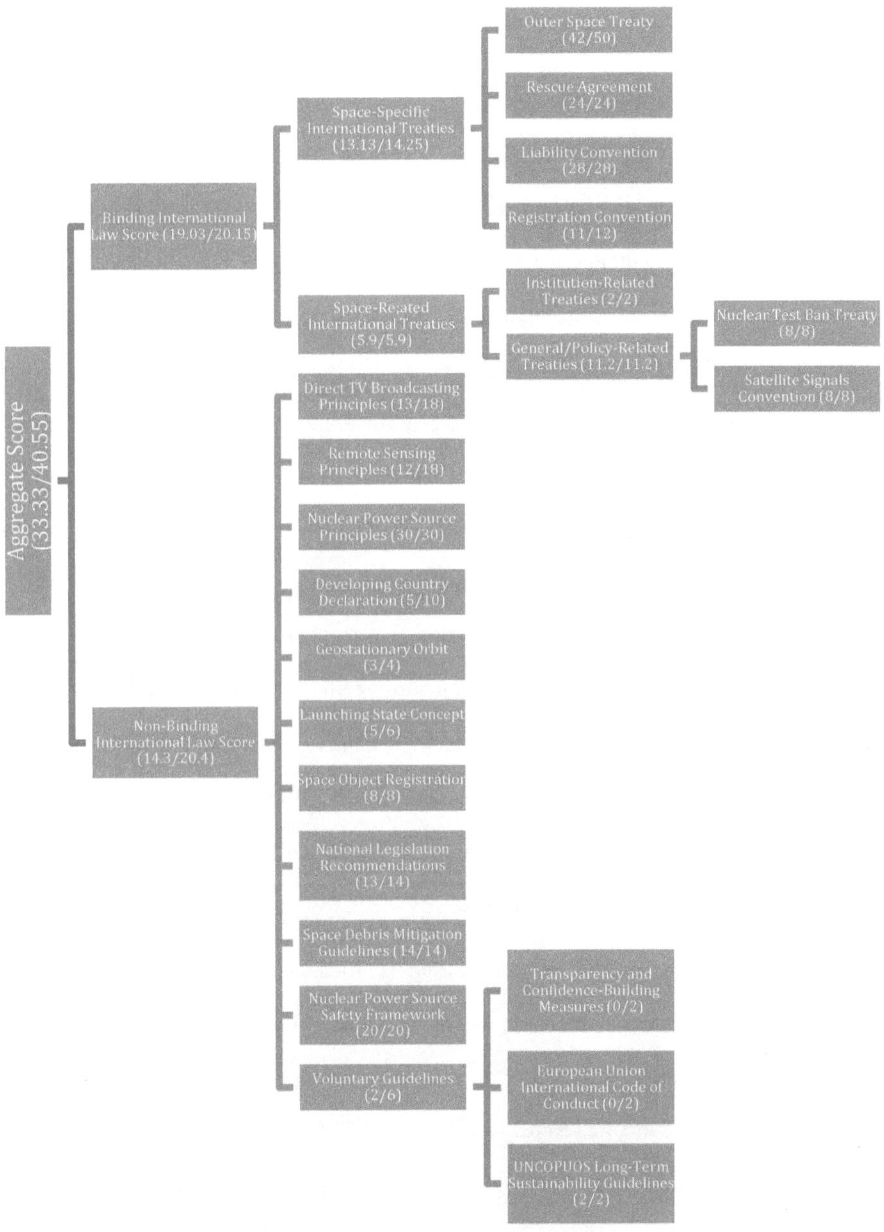

5.2.2 Scoring of South Korea

South Korea received an overall score of 30.80 out of a total possible score of 39.30. Based on the ratings ranges presented above, this score suggests South Korea is a mostly compliant country in adhering to international space law and norms. Specifically, South Korea received a score of 18.28 out of a possible score of 18.90 for the binding international law category, and it received a score of 12.40 out of a possible score of 20.40 for the non-binding international law category. Similar to the United States, South Korea performs well in terms of ratifying and adhering to formal treaties and their various provisions, but it falters somewhat as it relates to the non-binding UN principles, resolutions, and guidelines that promote greater normative behaviour in space. South Korea's performance in this regard may be due to its relatively more recent entry in to space activities, and as such has not rapidly enough developed laws or other structures to formalize many of these non-biding norms. For example, this is observed in the country's adherence to the Space Debris Mitigation Guidelines, where the country lacks a formal legal requirement for space activities to always comply with the guidelines. Similar to the United States, South Korea also falters on the remote sensing principles as it conducts its own satellite surveillance without full respect of other states' sovereignty, in particular North Korea. Progress in complying with these areas of the framework could lead to an improved rating in future assessments of South Korea.

Figure 5.5 below presents the scoring of South Korea for the various elements comprising the assessment framework, showing the assessed score out of the potential score possible for each element. Appendix II includes the detailed scoring for each question within the framework applied to South Korea.

Figure 5.5: Scoring of Assessment Framework Elements for South Korea

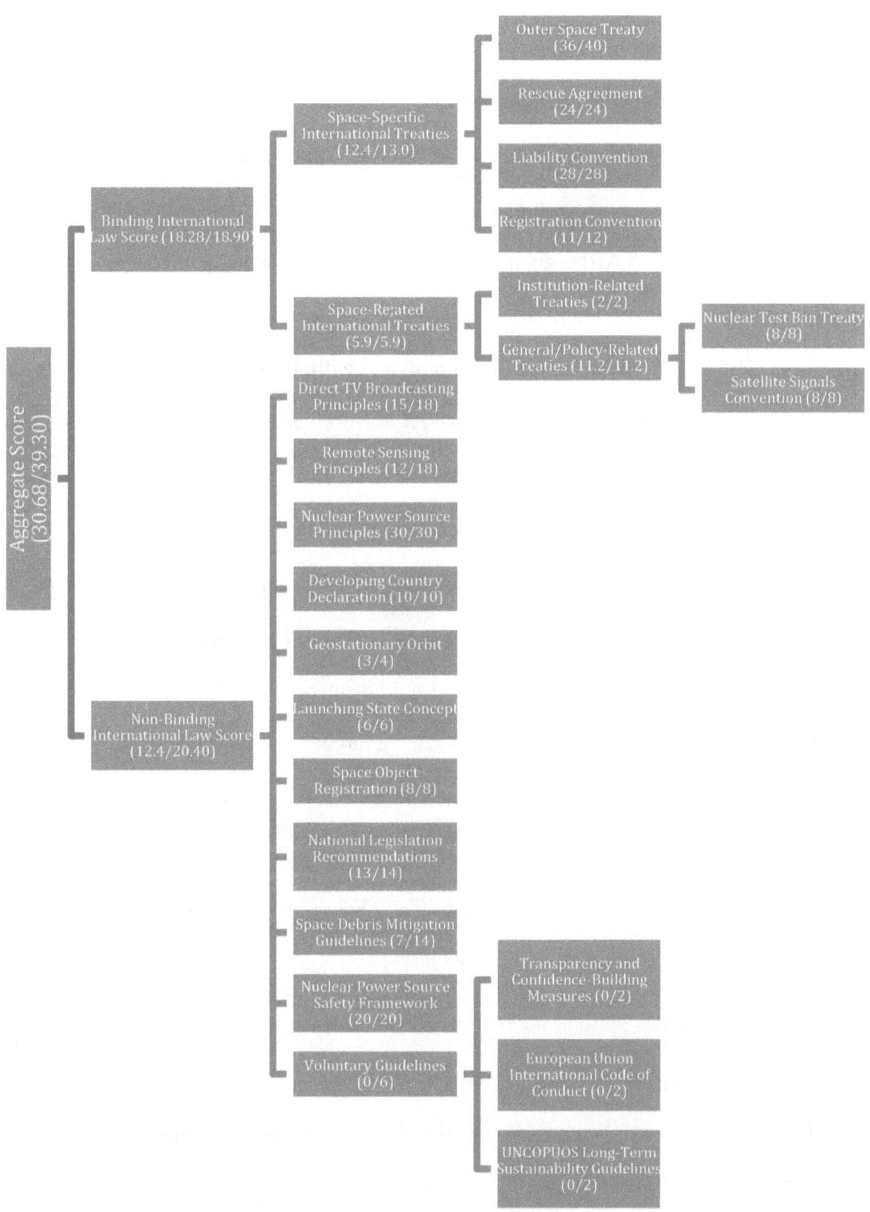

5.2.3 Scoring of South Africa

South Africa received an overall score of 19.80 out of a total possible score of 30.70, suggesting the country is mostly compliant in adhering to international space law and norms, although it sits on the cusp of the partially compliant rating and risks slipping in to that rating if it does not make improvements. As an emerging spacefaring country, many framework questions were not assessed toward South Africa given its lower level of space activity as compared to South Korea and the United States. Nonetheless, the framework demonstrates that South Africa does have room for improvement in further complying with international space and norms, particularly in the non-binding category. South Africa received a score of 12.98 out a possible score of 16.90 for the binding international law category, and it received a score of 6.40 out of a possible score of 13.60 for the non-binding category. South Africa's lacklustre performance in the binding international law category is due in part to the country not having registered its surveillance satellite Kondor-E, which was reported separately by the Russian Federation under a filing with the UN when it launched the satellite on South Africa's behalf. Improving transparency of its space operations, which can be argued for most if not all spacefaring countries, can significantly increase the country's score in future assessments. The country's poor performance in the non-binding international law category is largely due to its legal and administrative frameworks failing to formally incorporate proactive guidelines, such as those around space debris mitigation, in their procedures.

Figure 5.6 below presents the scoring of South Africa the various elements comprising the assessment framework, showing the assessed score out of the potential score possible for each element. Appendix III includes the detailed scoring for each question within the framework applied to South Africa.

Figure 5.6: Scoring of Assessment Framework Elements for South Africa

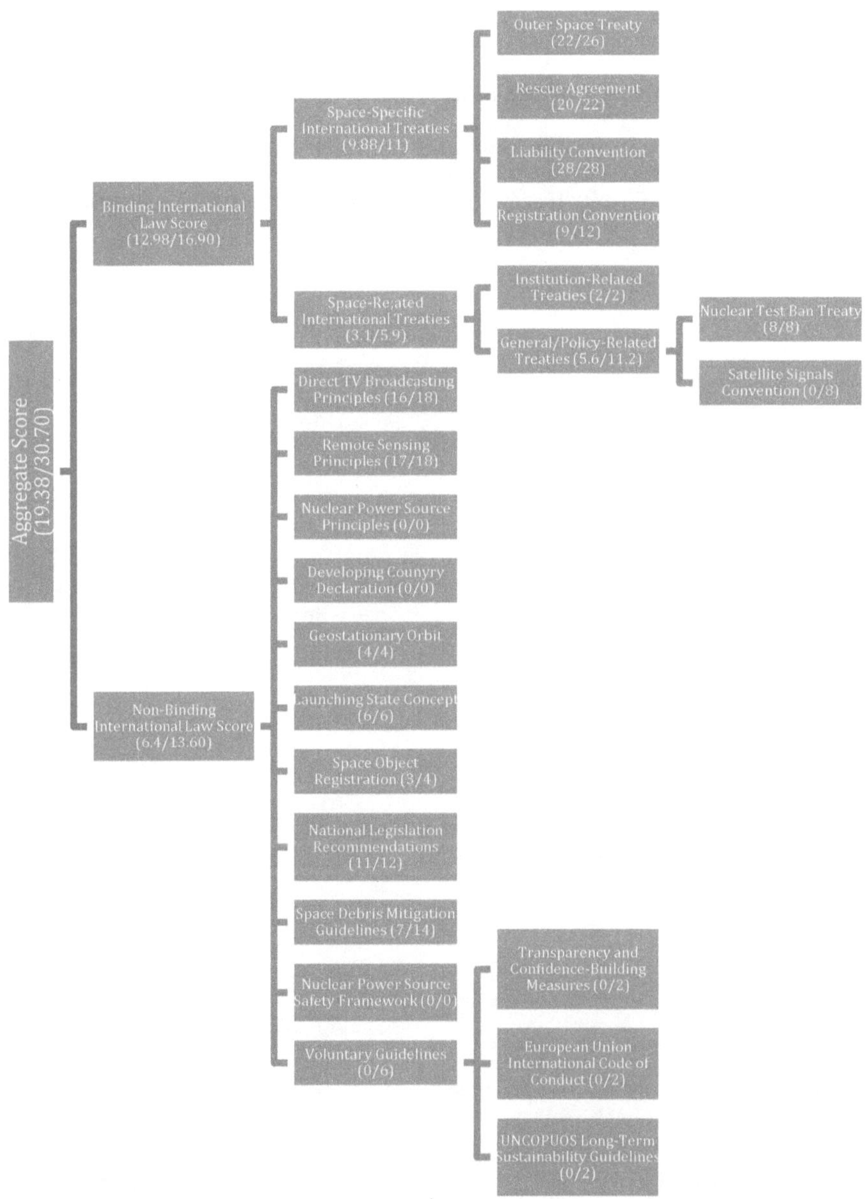

Now that the framework and its associated questions and scoring methods have been described above, including rationales for the inclusion and exclusion of particular articles and provisions of international space law treaties, principles, resolutions, and guidelines as identified by the UN Office of Outer Space Affairs, this book will conclude in the next chapter by discussing the scoring ramifications in addition to a discussion on the applicability of such a compliance assessment framework for advocacy and policymaking purposes.

6 CONCLUSION

We have developed and presented a scoring framework to assess the compliance of states with international law and norms pertaining to outer space activities. The framework accounts for the different levels of space activities and capabilities in advanced, intermediate, and emerging spacefaring countries. The scoring of each relevant article and provision within treaties, principles, resolutions, and guidelines helps to illuminate the process through which document-level scores, subsection scores, category scores, and total scores would be calculated using the framework for assessing space law compliance. The scoring also demonstrates the various scenarios in which scores can be calculated, including the range of what scores are possible using the assessment questions. The highest score that could be achieved by a country is 40.55 points for an advanced country, 39.30 for an intermediate country, and 30.70 for an emerging country, whereas the lowest score that could be achieved by a country in any tier is 0 points.

In order to ensure correct weighting and consideration of the importance of binding international law found in treaty documents versus non-binding norms set forth in principles, resolutions, and guidelines, the scoring of questions within each subsection had to be calculated using a different range of values possible for each question. For example, the framework questions are divided into two categories: a binding international law category and a non-binding international principles, resolutions, and guidelines category.

The binding international law category was further divided into two subsections: a subsection dealing only with space-specific international treaties (these are the four out of the five commonly known space treaties, excluding the Moon Agreement for reasons described previously) and a subsection dealing with space-related international treaties. In order to assign more weight to the space-specific treaties to account for their exclusive focus on the exploration and use of outer space and the principles that are affirmed throughout these documents, the space-specific treaties subsection was weighted 70 percent and the space-related treaties subsection was weighted 30 percent of the total category score. Irrespective of the subsection, each question appearing within this category on binding international law was able to be scored on a scale of values ranging from two (2) for most compliant to zero (0) for least compliant. The only exception to this scoring approach within the binding international law category pertains to the scoring of countries participating in international institutions, which was scored on a range of two (2) for countries that participate or had participated in at least two international institutions set forth by treaties to zero (0) for those countries that had not participated in any treaty-based institutions. The scoring for this particular question was prescribed in such a manner so as to recognise the importance of the principle of international cooperation in outer space policy and activities without unduly punishing a country by assigning a zero (0) score for having not participated in what were region-specific (for example, EUTELSAT and ARABSAT) or at the time considered politically-aligned institutions (for example, ITSO was considered to be Western-oriented institution and INTERSPUTNIK was considered to be a Soviet-oriented institution during the Cold War). The same approach was used in the non-binding law category to determine countries' voluntary levels of subscription to guidelines on transparency and confidence-building, the European Union code of conduct, and the UNCOPUOS long-term space sustainability guidelines.

Given the scoring of questions for the binding international law category was done on a range of two (2) to zero (0), the scoring of questions for the non-binding principles, resolutions, and guidelines was done using the same possible range of scores. As described above, this was intentional to weight the more significant and consequential of the two sets of questions and those most forceful in terms of driving policy among countries. Unlike the binding international categories, which had weighted subsections, the non-binding category does not weight specific principles, resolutions, or guidelines more heavily than others, therefore all questions within this category are assigned equal weight for arriving at the category score.

Once each category's score is assessed, both scores are summed to produce the total score for the country. Given the range of possible total scores is from 40.55 to 0 points for an advanced spacefaring country, the scores can be categorised according to intervals of 8.11 points. Similarly, for an intermediate country, an interval of 7.86 points can be used to categorise scores ranging from 39.30 to 0, and an interval of 6.14 points can be used to categorise scores ranging from 30.70 to 0 for an emerging country.

6.1 Interpretation of assessment results

As described above, the framework produces five broad yet useful categories for understanding the general level of compliance a given country follows with regard to international space law and norms. A country that is highly compliant follows all or most of international space law and norms in both policy and practice, whereas a mostly compliant country follows most international space law and norms but has room to improve in both policy and practice. As demonstrated in the application of the framework to the United States, South Korea, and South Africa, highly compliant and mostly compliant countries tend to perform better in the binding international law category while lacking in the non-binding category for principles, resolutions, and guidelines.

A partially compliant country, falling within the middle of the possible ratings, likely follows most of the binding international law with some inconsistencies, while lacking most adherence to the non-binding law and norms. A rating of minimally compliant and a rating of non-compliant indicate that a country does not observe international space law and norms, and is likely a threat to outer space peace, safety, security, and sustainability.

The approach proposed within this book, as previously discussed, falls within two broader movements, the evidence-based programming and policymaking movement and the ratings and rankings movement. Therefore, the value of this framework rests with its ability to contribute in a meaningful manner to informed space policy dialogue and policymaking, in addition to making space policy and practices generally accessible and consumable by a larger public audience for education and awareness-raising purposes. The framework intends to assist academics and policymakers to identify countries needing improvement in their space policies and locating the specific issue areas needed to make those improvements. Moreover, the framework, and the results it produces, intends to aid advocates, civil society, and the broader public to identify countries and issue areas in which they could exert pressure to encourage reform and best policy and practice in compliance with international law and norms. Finally, the framework intends to serve as a foreign policy and diplomatic tool whereby leading compliant countries can use the results of the ranking to exert pressure on their peers to improve compliance.

This approach subscribes to a similar logic found within the evidence-based programming and policymaking movement in that it assumes that increased data and analysis will lead to improved policymaking. This approach further assumes that improved policymaking and compliance with international space law and norms will lead to fulfilment of the objectives set forth within international space policy documents, including the achievement and maintenance of order, peace, safety, security, and sustainability in outer space.

The application of the framework on an annual basis also offers useful information and analysis that allows for the tracking of change over time, including improvement or worsening in compliance. This would assist academics, policymakers, and civil society and the public alike to, in near-real time, identify issue areas and undertake studies for understanding why such improvements or declines occur and to campaign for changes within the issue area of concern. With continued application over time, this framework would further normalise and propagate the policy and practices captured within the international space law and norms documents, and assist new space actors, including emerging spacefaring countries, in easily identifying and adopting leading best policy and practice.

6.2 Model application and use of the framework

Beyond the application of the framework to the three case studies within this paper, the real-world application and use of the framework must be considered. As such, questions of who performs the assessment and to whom it applies with what consequences arise. Although it remains outside the scope of this book to determine who will perform the application of the proposed framework, it is envisioned that the framework will be applied by an independent group of civil society organisations and academic institutions in anticipation that these groups would not be as biased or predisposed to certain ratings and assessments as would be corporate or governmental entities. The most successful of the ranking and rating assessments as described previously are those that have the most claim to independence and lack of interference from those groups with an interest to achieving a certain outcome. Indeed, the groups implementing these assessments have stressed the importance of receiving support from private and non-governmental sources to avoid the impression of undue influence being exerted on outcomes of the assessments. Therefore, individual corporations and governmental agencies would likely form the least likely candidates constituting a group overseeing the application of the framework proposed herein given their vested interests in obtaining a favourable rating. Existing space civil society groups and relevant and qualified academic institutions, or centres or parts thereof, would comprise the most likely set of candidates responsible for application and implementation of the framework.

Given the group of organisations, or a representative body comprising representatives of such groups, would likely bear responsibility for applying the framework and arriving at its assessment conclusions ranging from highly compliant to non-compliant, the consequences of such an assessment would largely be unenforceable in the strictest sense of the word. International law itself is largely unenforceable unless states within the international community agree to subscribe to following a particular enforcement action. The consequences of a low assessment rating for a specific country would thus be exclusion from opportunities for collaboration with other spacefaring countries, and other 'soft' forms of consequences resulting from not observing binding and non-binding laws and norms accepted by the international space community. Similar to those ranking and rating assessments described previously, the consequences of a low ranking or rating are not implemented by a single government or international governmental organisation but are rather the opportunity costs and negative attention those low performers encounter. For example, the inclusion of Freedom House's *Freedom in the World* rating in the Millennium Challenge Corporation's decision-making process on to whether award a development loan to a country depends of their rating in the assessment. While Freedom House as the civil society organisation is responsible for the assessment, the consequences of the rating are evidenced in the use of the rating in other governmental or international governmental processes, such as loan decision-making, thus resulting in lost investment and opportunity costs for those who are rated poorly. It is envisioned the results of the application of this framework would operate similarly. For example, poorly rated countries would lose out on the opportunity to collaborate with other spacefaring countries, thus resulting in missed opportunities to learn new knowledge and technical expertise that would have otherwise been afforded in multilateral space mission collaborations.

As the framework contributes to a norm-setting process across the international community, countries that perform well in the assessment would want to work with other well-performing countries, knowing they share the same or similar approaches to the access and use of outer space and compliance with binding and non-binding international space law and norms. Well-performing countries, in turn, would be less willing to work with poor performers as they may introduce elements of risk and non-compliance in joint space activities. Therefore, the consequences of inadequate compliance with international space law and norms would be imposed by the community of spacefaring nations themselves as positive behaviour would be encouraged and negative behaviour discouraged by the framework. To reinforce this approach, the example previously cited on the country of Georgia improving its *Ease of Doing Business* rating yielded significant returns and increases in foreign direct investment after it advanced its rating.

As such, the framework only functions when applied to state actors as state actors, according to international space law, are ultimately responsible for licensing and supervision of space activities, irrespective of whether they are conducted by national public or private entities. The regulatory and licensing and supervisory environment a state actor creates within its domain ultimately affects the character and nature of behaviours public and private entities follow. A government that remains lax in its licensing and supervision requirements would thus consequently create a lax and, likely, non-compliant environment in which private actors alongside public actors would operate. Therefore, the framework would be intended to be applied primarily to state actors as the ultimate responsible parties for licensing and supervision of space activities.

6.3 Recommendations and refinement of the framework

The approach described herein does not claim to completely address all areas of space policy and practice given that, as mentioned above, new governmental and non-governmental entrants continue to increasingly participate in outer space activities and that new technologies and growing markets are evolving rapidly, posing new challenges and issues areas for outer space policymaking. For example, the issue of space debris has recently come to the forefront of space policy challenges, whereas as recently as three decades ago it was not considered a pressing issue. Therefore, it remains incumbent upon this framework to prove flexible and adaptable, incorporating new space law and norms for evaluation of compliance as they emerge at the international level. This demonstrates the framework's cognizance of the fact that the space policy domain is continually changing and evolving as new actors and technological and economic factors facilitate increased access to, and use of, outer space.

As such, this book concludes by inviting additional review of the proposed framework and suggestions for improvement so as to ensure its responsiveness and usefulness to the needs of the international space community. Purposeful decisions have been taken in terms of the framework's structure and the composition, such as the exclusion of rating the Moon Agreement as an example, or the weighting of binding law versus non-binding norms contained within principles, resolutions, and guidelines. If it is determined as time evolves that these decisions are not responsive to the needs of the policy and civil society communities for purposes of promoting improved compliance, and therefore promoting the peace, security, and sustainability of outer space, then the framework remains adjustable and open to incorporating additional components for assessment. The book also invites further review for strengthening the methodological soundness of the assessment, and for consideration of its applicability to new space entrants.

Finally, this book suggests as next steps in a future research project the application of the proposed framework to a small group of established and current spacefaring countries. This would aid the assessment for identifying areas for improvement, and perhaps suggest a revision and further detailing of the five broad compliance categories established here (i.e. highly compliant, mostly compliant, partially compliant, minimally compliant, and non-compliant) in order to allow greater differentiation of compliance among space actors. Beyond this initial application of the framework, the assessment could then continue to expand the group of countries it measures to ultimately include all current spacefaring countries and measure them thereafter on a continuous, annual basis.

REFERENCES

Broome, Andre and Quirk, Joel, 2015, "Governing the world at a distance: the practice of global benchmarking," *Review of International Studies*, Vol 41 (5), Accessed via: https://www.cambridge.org/core/journals/review-of-international-studies/article/governing-the-world-at-a-distance-the-practice-of-global-benchmarking/06B12DE43C6C5D647B701206D6B93826

Burnell, Peter and Calvert, Peter, 2004, *Civil Society in Democratization*, Frank Cass and Company Limited, London, Accessed via: https://ebookcentral.proquest.com/lib/aul/reader.action?docID=183204&query

Bush, Sarah, 2017, "Should we trust democracy ratings? New research finds hidden biases," *The Washington Post*, Accessed via: https://www.washingtonpost.com/news/monkey-cage/wp/2017/11/07/why-do-we-trust-certain-democracy-ratings-new-research-explains-hidden-biases/?utm_term=.547fbca98863

CIVICUS, 2018, "CIVICUS Monitory Methodology Paper," CIVICUS, Accessed via: https://www.civicus.org/documents/civicus-monitor-methodology-paper.pdf

CIVICUS, 2018, "How it Works," CIVICUS, Accessed via: https://monitor.civicus.org/about/

Cooley, Alexander and Snyder, Jack, 2016, *Ranking the World: Grading States as a Tool of Global Governance*, Cambridge University Press

Cornish, Flora, 2015, "Evidence synthesis in international development: a critique of systematic reviews and a pragmatist alternative," *Anthropology and Medicine*, Vol. 22 (3), Accessed via: https://www.ncbi.nlm.nih.gov/pmc/articles/PMC4960511/

Engle Merry, Sally, Davis, Kevin and Kingsbury, Benedict, 2015, *The Quiet Power of Indicators: Measuring Governance, Corruption, and Rule of Law*, Cambridge Studies in Law and Society, Cambridge University Press

Ferrarello, Molli, 2017, "What "America First" means for US foreign aid," The Brookings Institution, Accessed via: https://www.brookings.edu/blog/brookings-now/2017/07/27/what-america-first-means-for-us-foreign-aid/

Foss Hansen, Hanne, 2014, "Organisation of evidence-based knowledge production: Evidence hierarchies and evidence typologies," *Scandinavian Journal of Public Health*, Vol 42 (13), Accessed via: http://journals.sagepub.com/doi/full/10.1177/1403494813516715

References

Freedom House, 2018, "Freedom in the World 2018: Methodology," Freedom House, Accessed via:

https://freedomhouse.org/report/methodology-freedom-world-2018

Gisselquist, Rachel, 2014, "Developing and evaluation governance indexes: 10 questions," *Journal of Policy Studies*, Vol 34 (5), Accessed via: https://www.tandfonline.com/doi/full/10.1080/01442872.2014.946484

Gregory, Robert, 2014, "Assessing 'good governance': 'scientific' measurement and political discourse," *Policy Quarterly*, Vol 10 (1), Accessed via: https://ojs.victoria.ac.nz/pq/article/view/4478

International Telecommunications Union, 2018, "Position of Member States in relation to the Acts of the Union," International Telecommunications Union, Accessed via:

https://www.itu.int/en/membership/Pages/member-states-position.aspx

Jones, Harry, 2012, "Promoting evidence-based decision-making in development agencies," Overseas Development Institute, Accessed via: https://www.odi.org/sites/odi.org.uk/files/odi-assets/publications-opinion-files/7575.pdf

Kelley, Judith and Simmons, Beth, 2014, "Politics by Number: Indicators as Social Pressure in International Relations," *American Journal of Political Science*, Vol 59 (1), Accessed via:

https://onlinelibrary.wiley.com/doi/abs/10.1111/ajps.12119

Kelley, Judith, 2017, "The State Department just released its human trafficking report. Here's why it matters," *The Washington Post*, Accessed via: https://www.washingtonpost.com/news/monkey-cage/wp/2017/07/03/the-state-department-just-released-its-trafficking-in-persons-report-heres-why-that-matters/?utm_term=.be9cfaaa5129

Kilibarda, Pavle, 2017, "Space law revisited : The regime of international liability in space," *Humanitarian Law and Policy*, International Committee of the Red Cross, Accessed via: https://medium.com/law-and-policy/space-law-revisited-the-regime-of-international-liability-in-space-66a864fa5157

Kim, Doo Hwan, 2012, "Space Policy and Law in the Republic of Korea," United Nations Office of Outer Space Affairs, Accessed via: http://www.unoosa.org/pdf/pres/2010/SLW2010/02-09.pdf

Kim, Felix, 2018, "South Korea to strengthen reconnaissance satellite capabilities," Indo-Pacific Defense Forum, Accessed via: http://apdf-magazine.com/south-korea-to-strengthen-reconnaissance-satellite-capabilities/

Kleinman, Matthew, 2013, "SpaceLaw 101: An Introduction to Space Law," American Bar Association, Accessed via:

References

https://www.americanbar.org/groups/young_lawyers/publications/the_101_201_practice_series/space_law_10
1_an_introduction_to_space_law/

Kraay, Aart, Kaufmann , Daniel and Mastruzzi, Massimo, 2010, "The worldwide governance indicators:
methodology and analytical issues," World Bank Group, Accessed via:
https://elibrary.worldbank.org/doi/abs/10.1596/1813-9450-5430

Krisch, Nico, 2005, "International Law in Times of Hegemony: Unequal Power and the Shaping of the
International Legal Order," *The European Journal of International Law*, Vol 16 (3), Accessed via:
http://ejil.org/pdfs/16/3/301.pdf

Legatum Institute, 2017, "The Legatum Prosperity Index 2017: Methodology Report," Legatum Institute,
Accessed via: https://prosperitysite.s3-
accelerate.amazonaws.com/9615/1186/6075/Legatum_Prosperity_Index_2017_Methodology_Report.pdf

Majaja, Nomfuneko, 2018, "Space Related Issues," Portfolio Committee on Trade and Industry, South Africa
Department of Trade and Industry, Accessed via:
https://www.thedti.gov.za/parliament/2018/Space_Issues.pdf

Majaja, Nomfuneko and Maruping, Ponthso, 2017, "Repeal of the Space Affairs Act No. 84 of 1993 by the
South Africa Outerspace Bill," Portfolio Committee on Trade and Industry, South Africa Department of Trade
and Industry, Accessed via: https://www.thedti.gov.za/parliament/2017/Space_Draft_legislation.pdf

Martinez, Peter, 2016, "The development and initial implementation of South Africa's national space policy,"
Space Policy, Vol 37 Part 1, Accessed via:
https://www.sciencedirect.com/science/article/pii/S026596461630087X

Michener, Gregory, 2015, "Policy Evaluation via Composite Indexes: Qualitative Lessons from International
Transparency Policy Indexes," *World Development*, Vol 74, Accessed via:
https://www.sciencedirect.com/science/article/pii/S0305750X1500100X

Millennium Challenge Corporation, 2017, "Report on the Criteria and Methodology for Determining the
Eligibility of Candidate Countries for Millennium Challenge Account Assistance for Fiscal Year 2018,"
Millennium Challenge Corporation, Accessed via: https://www.mcc.gov/resources/doc/report-selection-
criteria-methodology-fy18

References

Mo Ibrahim Foundation, 2017, "2017 Ibrahim Index of African Governance: Index Report," Mo Ibrahim
Foundation, Accessed via: http://s.mo.ibrahim.foundation/u/2017/11/21165610/2017-IIAG-
Report.pdf?_ga=2.233719902.1566231500.1534030003-1375651803.1534030003#page=15

Mo Ibrahim Foundation, 2018, "Ibrahim Index of African Governance (IIAG)," Mo Ibrahim Foundation,
Accessed via: https://mo.ibrahim.foundation/iiag/

Mo Ibrahim Foundation, 2018, "Methodology," Mo Ibrahim Foundation, Accessed via:
https://mo.ibrahim.foundation/iiag/methodology/

National Aeronautics and Space Administration, "U.S. Government Orbital Debris Mitigation Standard
Practices," National Aeronautics and Space Administration, Accessed via:
https://orbitaldebris.jsc.nasa.gov/library/usg_od_standard_practices.pdf

Nelson Espeland, Wendy and Sauder, Michael, 2007, "Rankings and Reactivity: How Public Measures Recreate
Social Worlds," *American Journal of Sociology*, Vol 113 (1), Accessed via: https://www.stmarys-
ca.edu/sites/default/files/attachments/files/rankings-and-reactivity-2007.pdf

Oduntan, Gbenga, 2003, "The Never Ending Dispute: Legal Theories on the Spatial Demarcation Boundary
Plane between Airspace and Outer Space," Hartfordshire Law Journal, Vol 1 (2), Accessed via:
https://pdfs.semanticscholar.org/d81b/b88753eaced1ec97cc56baed8cc25ffdc9d6.pdf

Petros, Eloi, 2018, "Space Debris Mitigations in National Space Legislation," Institute for Space and
Telecommunications Law, Accessed via:
https://indico.esa.int/event/128/attachments/728/796/Space_Debris_mitigations_in_national_space_legislation
.pdf

Project Ploughshares, 2018, "Space Security Index," Project Ploughshares, Accessed via:
http://ploughshares.ca/space-security/space-security-index/

Radio Free Asia, 2015, "TV Censors Can't Stop North Koreans From Watching South Korean Programs,"
Radio Free Asia, Accessed via: https://www.rfa.org/english/news/korea/tv-censors-cant-stop-north-koreans-
from-watching-south-korean-programs-06052015160410.html

Secure World Foundation, 2018, "Space Sustainability: A Practical Guide," Secure World Foundation,
Broomfield, Colorado, United States

South African Council for Space Affairs, 2018, "National Registry of Objects Launched into Outer Space,"
South Africa Department of Trade and Industry, Accessed via: http://www.sacsa.gov.za/registry/

Space Security Index, 2017, "Space Security Index 2017," Space Security Index

References

Space Security Index, 2018, "Space Security Index 2018: Executive Summary," Space Security Index, Accessed via: http://spacesecurityindex.org/wp-content/uploads/2018/06/SSIExecutiveSummary2018.pdf

Space Security Index, 2018, "SSI: Space Security Index," Space Security Index, Accessed via: http://spacesecurityindex.org/

Space Security Index, 2014, "The Space Security Index Factsheet," Space Security Index, Accessed via: http://spacesecurityindex.org/wp-content/uploads/2014/10/spacesecurityindexfactsheet.pdf

Steiner, Nils, 2016, "Comparing Freedom House Democracy Scores to Alternative Indices and Testing for Political Bias: Are US Allies Rated as More Democratic by Freedom House?" *Journal of Comparative Policy Analysis*, Vol 18 (4), Accessed via: https://www-tandfonline-com.proxyau.wrlc.org/doi/pdf/10.1080/13876988.2013.877676?needAccess=true

Sunn Bush, Sarah, 2017, "The Politics of Rating Freedom: Ideological Affinity, Private Authority, and the Freedom in the World Ratings," *American Political Science Association*, Vol 15 (3), Accessed via: https://sites.temple.edu/sarahbush/files/2012/08/Bush-POP-2017.pdf

Tars, Eric and Bhattarai, Deodonne, 2011, "Opening the Door to the Human Right to Housing: The Universal Periodic Review and Strategic Federal Adovcacy for a Rights-Based Approach to Housing," *Clearing House Rev. 197*, Vol 45, Accessed via: https://heinonline.org/HOL/Page?collection=journals&handle=hein.journals/clear45&id=208

The Heritage Foundation, 2018, "2018 Index of Economic Freedom: About the Index," The Heritage Foundation, Accessed via: https://www.heritage.org/index/about

The Heritage Foundation, 2018, "2018 Index of Economic Freedom: Interactive Heat Map," The Heritage Foundation, Accessed via: https://www.heritage.org/index/heatmap

Transparency International, 2018, "Corruption Perceptions Index 2017: Full Source Description," Transparency International, Accessed via: https://www.transparency.org/news/feature/corruption_perceptions_index_2017

Transparency International, 2018, "Corruption Perceptions Index 2017: Technical Methodology Note," Transparency International, Accessed via: https://www.transparency.org/news/feature/corruption_perceptions_index_2017

United Nations Development Group, 2018, "Theory of Change: UNDAF Companion Guidance," United Nations, Accessed via: https://undg.org/wp-content/uploads/2017/06/UNDG-UNDAF-Companion-Pieces-7-Theory-of-Change.pdf

References

United Nations Education, Scientific and Cultural Organization, 2017, "More about the nature and status of the legal instruments and programmes," United Nations, Accessed via: http://www.unesco.org/new/en/social-and-human-sciences/themes/advancement/networks/larno/legal-instruments/nature-and-status/

United Nations Human Rights Council, 2018, "Basic facts about the UPR," United Nations, Accessed via: https://www.ohchr.org/EN/HRBodies/UPR/Pages/BasicFacts.aspx

United Nations Office for Outer Space Affairs, 2015, "Information furnished in conformity with the Convention on Registration of Objects Launched into Outer Space," United Nations, Accessed via: http://www.unoosa.org/oosa/oosadoc/data/documents/2015/aac.105/aac.1051095_0.html

United Nations Office for Outer Space Affairs, 2017, "International Space Law: United Nations Instruments," United Nations, Accessed via: http://www.unoosa.org/res/oosadoc/data/documents/2017/stspace/stspace61rev_2_0_html/V1605998-ENGLISH.pdf

United Nations Office for Outer Space Affairs, 2018, "Selected Examples of National Laws Governing Space Activities: South Africa," United Nations, Accessed via: http://www.unoosa.org/oosa/en/ourwork/spacelaw/nationalspacelaw/south_africa/space_affairs_amendment_act_1995E.html

United Nations Office for Outer Space Affairs, 2018, "Selected Examples of National Laws Governing Space Activities: Republic of Korea," United Nations, Accessed via: http://www.unoosa.org/oosa/en/ourwork/spacelaw/nationalspacelaw/republic_of_korea/space_development_promotions_actE.html

United Nations, 2018, "Agreement governing the Activities of States on the Moon and Other Celestial Bodies," United Nations Treaty Collection, Accessed via: https://treaties.un.org/pages/ViewDetails.aspx?src=TREATY&mtdsg_no=XXIV-2&chapter=24&clang=_en

United States Code, "National and international orbital debris mitigation," 42 United States Code Section 18441, Accessed via: https://www.law.cornell.edu/uscode/text/42/18441

UPR-Info, 2018, "Role of CSOs," UPR-Info, Accessed via: https://www.upr-info.org/en/how-to/role-ngos

Valters, Craig, 2014, "Theories of Change in International Development: Communication, Learning, or Accountability?" The Justice and Security Research Programme, The Asia Foundation, Accessed via: http://www.lse.ac.uk/internationalDevelopment/research/JSRP/downloads/JSRP17.Valters.pdf

Vogel, Isabel, 2012, "Review of the use of 'Theory of Change' in international development," UK Department

of International Development, Accessed via:

http://www.theoryofchange.org/pdf/DFID_ToC_Review_VogelV7.pdf

Wally, Nermine, 2018, "Sharing Development Pathways – On Evaluation Theory and Practice," Cairo Institute

of Liberal Arts and Sciences, Accessed via: http://www.ci-las.org/monitoring-and-evaluation.html

World Bank, 2018, "Distance to Frontier and Ease of Doing Business Ranking," World Bank, Accessed via:

http://www.doingbusiness.org/~/media/WBG/DoingBusiness/Documents/Annual-Reports/English/DB18-

Chapters/DB18-DTF-and-DBRankings.pdf

World Bank, 2018, "Worldwide Governance Indicators," World Bank, Accessed via:

http://info.worldbank.org/governance/wgi/#doc

World Intellectual Property Organisation, 2018 "Brussels Convention," World Intellectual Property

Organisation, Accessed via: https://www.wipo.int/treaties/en/ShowResults.jsp?lang=en&treaty_id=19

Zak, Anatoly, 2015, "Russia orbits South-Africa's first spy satellite Kondor-E," Russian Space Web, Accessed

via: http://www.russianspaceweb.com/kondor-e.html

APPENDICES

Appendix I:
Scoring for United States

A	Aggregate Score
Binding International Law Score *(Enter total from last line of Table B.1)*	Advanced: 19.03 / 20.15
Non-Binding International Law Score *(Enter total from last line of Table B.2)*	Advanced: 14.30 / 20.40
Total *(Sum binding and non-binding international law scores)*	Advanced: 33.33 / 40.50

Appendix I: Scoring for United States

B.1	Binding International Law Score
Space-specific International Treaties *(Use total from last line of Table B.1.1 and multiply by .50)*	Advanced: 13.13 / 14.25
Space-related International Treaties *(Use total from last line of Table B.1.2 and multiply by .50)*	Advanced: 5.90 / 5.90
Total *(Sum space-specific and space-related treaties scores and then bring total to Table A)*	Advanced: 19.03 / 20.15

B.1.1	Space-specific International Treaties	Score
1. Treaty on Principles Governing the Activities of States in the Exploration and Use of Outer Space, including the Moon and Other Celestial Bodies (25 percent of section score)		
a. Has the country ratified or acceded to this treaty?	Yes, fully = 2 Yes, with reservations = 1 No = 0	2
b. Is the treaty domesticated in national laws or administrative regulations or procedures?	Full domestication = 2 Partial domestication = 1 No domestication = 0	2
c. Article I: Do country's laws or administrative regulations or procedures recognise the role of international law?	Full recognition = 2 Partial recognition = 1 No recognition = 0	2
d. Article I: Do the country's laws or administrative regulations or procedures affirm that the exploration and use of outer space should be carried out for the benefit and interests of all countries?	Full affirmation = 2 Partial affirmation = 1 No affirmation = 0	1
e. Article I: Do the country's laws or administrative regulations or procedures recognise outer space as the province of all mankind?	Full recognition = 2 Partial recognition = 1 No affirmation = 0	0
f. Article I: Do the country's laws or administrative regulations or procedures recognise that outer space should be free for exploration without discrimination and with free access to celestial bodies?	Full recognition = 2 Partial recognition = 1 No recognition = 0	2
g. Article I: Do the country's laws or administrative regulations or procedures recognise the freedom of scientific investigation in outer space?	Full recognition = 2 Partial recognition = 1 No recognition = 0	2
h. Article I: Does the country facilitate and participate in international scientific investigation?	Frequently = 2 Occassionally = 1 Never = 0	2
i. Article II: Do the country's laws or administrative regulations or procedures allow for, or encourage, the appropriation of outer space and celestial bodies by claims of sovereignty, use, or occupation?	Always = 0 Sometimes = 1 Never = 2	1
j. Article II: Does the country, by its actions or the actions of those within its jurisdiction, appropriate, or encourage the appropriation of, outer space and celestial bodies and resources by claims of sovereignty, use, or occupation?	Frequently = 0 Occassionally = 1 No = 2	1
Article III: Elements of this article are already captured in questions asked above, therefore no separate scoring is assessed.	N/A	
k. Article IV: Has the country placed into orbit or installed on celestial bodies weapons of mass destruction, including nuclear weapons?	Repeatedly = 0 Once = 1 No = 2	2
l. Article IV: Has the country established bases, installations, or fortifications, or tested weapons or conducted military manoeuvres on celestial bodies?	Repeatedly = 0 Once = 1 No = 2	2
m. Article IV: Do the country's laws or administrative regulations or procedures allow for the conduct of such types of activities as described in Question 'l' above?	Always = 0 Sometimes = 1 Never = 2	1
n. Article V: Has the country either provided assistance or rejected assistance to astronauts of other countries who are in distress?	Always provided / not applicable = 2 Sometimes provided = 1 Never provided = 0	2
o. Article V: Does the country proactively inform the United Nations and other countries regarding potentially dangerous phenomena to astronauts?	Always = 2 Sometimes = 1 Never = 0	2
p. Article VI: Does the country have in place an authorisation and supervision regime for space activities conducted by non-governmental entities?	Yes = 2 Partial = 1 No = 0	2
q. Article VII: Does the country recognise its liability as a launching state in its laws or administrative regulations or procedures?	Full recognition = 2 Partial recognition = 1 No recognition = 0	2
r. Article VII: Has the country denied its liability as a launching state for damage caused?	Always = 0 Sometimes = 1 Never = 2	2
s. Article VIII: Does the country maintain an updated national registry of launched space objects?	Always = 2 Sometimes = 1 Never = 0	2
t. Article VIII: Has the country ever returned a space object belonging to another country, but which landed in its jurisdiction?	Always / not applicable = 2 Sometimes = 1 Never = 0	2
u. Article IX: Do the country's laws or administrative regulations or procedures account for avoiding harmful contamination, including debris, of the space environment?	Full account = 2 Partial account = 1 No account = 0	2
v. Article IX: Do the country's laws or administrative regulations or procedures seek to limit damage to the Earth environment by extraterrestrial material?	Full account = 2 Partial account = 1 No account = 0	2
w. Article X: Does the country allow other countries to observe the flight of space objects that it launches?	Always = 2 Sometimes = 1 Never = 0	1
x. Article XI: Does the country inform the United Nations and other countries of its space activities, including the nature, conduct, location, and results of the activities?	Always = 2 Sometimes = 1 Never = 0	1
y. Article XII: Does the country allow access and visits to its outer space stations, installations, equipment, and vehicles, provided reasonable notification of a such visit is given by the visiting representatives?	Always / not applicable = 2 Sometimes = 1 Never = 0	2
Articles XIII, XIV, XV, XVI, and XVII: These are administrative articles, therefore no scoring is assessed.		
Treaty Score (Sum the above scores and multiply by .25)		10.5
2. Agreement on the Rescue of Astronauts, the Return of Astronauts and the Return of Objects Launched into Outer Space *(25 percent of section score)*		
	Yes, in full = 2	

Appendix I: Scoring for United States

a. Has the country ratified or acceded to this treaty?	Yes, with reservations = 1 No = 0		2
b. Is the treaty domesticated in national laws or administrative regulations or procedures?	Full domestication = 2 Partial domestication = 1 No domestication = 0		2
c. Article 1: Has the country notified the launching state(s) of astronauts who have suffered accidents, been in distress, or made emergency landings?	Always / not applicable = 2 Sometimes = 1 Never = 0		2
d. Article 1: Has the country notified the United Nations of astronauts who have suffered accidents, been in distress, or made emergency landings?	Always / not applicable = 2 Sometimes = 1 No = 0		2
e. Article 2: Has the country taken all possible steps to rescue and render all necessary assistance to astronauts who have suffered accidents, been in distress, or made emergency landings?	Always / not applicable = 2 Sometimes = 1 No = 0		2
f. Article 3: Has the country provided assistance in search and rescue operations for astronauts who have suffered accidents, been in distress, or made emergency landings?	Always / not applicable = 2 Sometimes = 1 No = 0		2
g. Article 4: Has the country safely and promptly returned to their launching states astronauts who have suffered accidents, been in distress, or made emergency landings?	Always / not applicable = 2 Sometimes = 1 No = 0		2
h. Article 5: Has the country notified the launching state and United Nations of a space object or part of a space object that has returned to Earth?	Always / not applicable = 2 Sometimes = 1 No = 0		2
i. Article 5: Has the country in which the returning space object landed practicably assisted to recover the object or its parts?	Always / not applicable = 2 Sometimes = 1 No = 0		2
j. Article 5: Has the country in which the returning space object landed returned the space object or its parts and provided their identifying data to the launching state?	Always / not applicable = 2 Sometimes = 1 No = 0		2
k. Article 5: Has the country in which the returning space object landed eliminated the possibility of its danger if it is found to be hazardous or deleterious?	Always / not applicable = 2 Sometimes = 1 No = 0		2
l. Article 5: Has the launching state paid for expenses related to the recovery and return of space object or its part incurred by the country in which the space object landed?	Always / not applicable = 2 Sometimes = 1 No = 0		2
Articles 6, 7, 8, 9, and 10: These are administrative articles, therefore no scoring is assessed.			
Treaty Score (Sum the above scores and multiply by .25)			6
3. Convention on International Liability for Damage Caused by Space Objects			
a. Has the country ratified or acceded to this treaty?	Yes, in full = 2 Yes, with reservations = 1 No = 0		2
b. Is the treaty domesticated in national laws or administrative regulations or procedures?	Full domestication = 2 Partial domestication = 1 No domestication = 0		2
Article I provides for definitions of terms used in the treaty, therefore no scoring is assessed.			
c. Article II: Do the country's laws or administrative regulations or procedures recognise its liability to pay compensation for damage caused by its space objects on the Earth's surface or aircraft in flight?	Full recognition = 2 Partial recognition = 1 No recognition = 0		2
d. Article II: Has the country refused to pay compensation in practice for damage caused by its space objects on the Earth's surface or aircraft in flight?	Always = 0 Sometimes = 1 Never / not applicable = 2		2
e. Article III: Do the country's laws or administrative regulations or procedures recognise its liability owing to its fault or fault of persons for which it is responsible in the case of damage to another space object, persons, or property in space of another launching state?	Full recognition = 2 Partial recognition = 1 No recognition = 0		2
f. Article III: Has the country refused to accept liability in the case of damage to another space object, persons, or property in space of another launching state caused by its fault or fault of persons for which it is responsible?	Always = 0 Sometimes = 1 Never / not applicable = 2		2
g. Article IV: Do the country's laws or administrative regulations or procedures recognise its absolute joint and several liability with a second-party launching state if its space object caused damage to the second-party launching state's space object, which then resulted in damage on the Earth's surface or aircraft in flight of a third state?	Full recognition = 2 Partial recognition = 1 No recognition = 0		2
h. Article IV: Does the country's laws or administrative regulations or procedures recognise its joint and several liability apportioned according to fault with a second-party launching state if its space object caused damage to the second-party launching state's space object, which then resulted in damage to a third state's space object, person, or property in space?	Full recognition = 2 Partial recognition = 1 No recognition = 0		2
i. Article IV: Has the country refused to pay compensation according to its fault in cases of joint and several liability for damage caused to a third-party state?	Always = 0 Sometimes = 1 Never / not applicable = 2		2
j. Article V: Do the country's laws or administrative regulations or procedures recognise joint and several liability in cases when it and another state jointly launch a space object?	Full recognition = 2 Partial recognition = 1 No recognition = 0		2
k. Article V: Do country's laws or administrative regulations or procedures account for its roles a joint launch participant if a space object is launched from its territory or facility?	Full account = 2 Partial account = 1 No account = 0		2
Articles VI and VII sets forth conditions under which other articles apply and is not actionable by states, therefore no scoring is assessed.			
l. Article VIII: Do the country's laws or administrative regulations or procedures provide for a claims and compensation process for damage which it may suffer by other launching states?	Full provision = 2 Partial provision = 1 No provision = 0		2
Article IX, X, XI, XII, and XIII prescribe the administrative details through which claims and compensation should be processed, therefore no scoring is assessed.			
Articles XIV, XV, XVI, XVII, and XVIII prescribe the administrative details through which a Claims Commission should be established and make adjudications in the case of no settlements through diplomatic channels established elsewhere in the treaty, therefore no scoring is assessed.			
m. Article XIX: Has the country refused or failed to pay compensation or an award decided by a Claims Commission as set forth in other treaty articles?	Always = 0 Sometimes = 1 Never / not applicable = 2		2
Article XX prescribes the process through which the expenses associated with the Claims Commission should be borne by state participants, therefore no scoring is assessed.			
n. Article XXI: Has the country rendered appropriate and rapid assistance in cases when a space object presents large-scale danger to human life or interferes with a population's living conditions or functioning of a vital centre?	Always / not applicable = 2 Sometimes = 1 Never = 0		2
Articles XXII, XXIII, XXIV, XXV, XXVI, XXVII, XXVIII: These are administrative articles, therefore no scoring is assessed.			
Treaty Score (Sum the above scores and multiply by .25)			7
4. Convention on Registration of Objects Launched into Outer Space			
a. Has the country ratified or acceded to this treaty?	Yes, in full = 2 Yes, with reservations = 1 No = 0		2

Appendix I: Scoring for United States

		Score
b. Is the treaty domesticated in national laws or administrative regulations or procedures?	Full domestication = 2 Partial domestication = 1 No domestication = 0	2
Article I provides for definitions of terms used in the treaty, therefore no scoring is assessed.		
c. Article II: Does the country maintain an updated national registry of launched space objects that it launches and that it launches jointly with other launching states?	Always / not applicable = 2 Sometimes = 1 Never = 0	2
Article III describes actions to be taken by the United Nations rather than a specific country, therefore no scoring is assessed.		
d. Article IV: Does the country provide detailed information to the United Nations for its launched space objects, including name of the country, space object registration number, date and territory of launch, basic orbital parameters, and general function of the space object?	Full information / not applicable= 2 Partial information = 1 No information = 0	1
e. Article IV: Does the country update the United Nations on space objects for which it previously shared information, but which are no longer in Earth orbit?	Always / not applicable = 2 Sometimes = 1 Never = 0	2
Article V describes a procedural administrative action, therefore no scoring is assessed.		
f. Article VI: Does the country assist with providing identifying information it has gained through space monitoring and tracking facilities for space objects that are unidentifiable and which are of a hazardous or deleterious nature?	Always = 2 Sometimes = 1 Never = 0	2
Articles VII, VIII, IX, X, and XII: These are administrative articles, therefore no scoring is assessed.		
Treaty Score (Sum the above scores and multiply by .25)		2.75
Total (Sum the above treaty score and bring total to Table B.1)		26.25

B.1.2	Space-related International Treaties	Score
a. Treaty Banning Nuclear Weapon Tests in the Atmosphere, in Outer Space and under Water		
i. Has the country ratified or acceded to this treaty?	Yes, in full = 2 Yes, with reservations = 1 No = 0	2
ii. Is the treaty domesticated in national laws or administrative regulations or procedures?	Full domestication = 2 Partial domestication = 1 No domestication = 0	2
iii. Article I: Do the country's laws or administrative regulations or procedures prohibit, prevent, and forbid the carrying out of nuclear weapon tests at places under its jurisdiction or control, including in the atmosphere and outer space and underwater?	Full prohibition = 2 Partial prohibition = 1 No prohibition = 0	2
iv. Article I: Since implementing the treaty, has the country carried out a nuclear weapon test at a place under its jurisdiction or control, including in the atmosphere and outer space and underwater?	Repeatedly = 0 Once = 1 Never = 2	2
Articles II, III, IV, and V: These are administrative articles, therefore no scoring is assessed.		
Treaty Score (Sum the above scores)		8
b. Convention Relating to the Distribution of Programme-Carrying Signals Transmitted by Satellite		
i. Has the country ratified or acceded to this treaty?	Yes, in full = 2 Yes, with reservations = 1 No = 0	2
ii. Is the treaty domesticated in national laws or administrative regulations or procedures?	Full domestication = 2 Partial domestication = 1 No domestication = 0	2
Article 1 provides for definitions of terms used in the treaty, therefore no scoring is assessed.		
iii. Article II: Do the country's laws or administrative regulations or procedures prevent the distribution on or from its territory programme-carrying signals by distributors for whom the emitted signal to or passing through the satellite is not intended?	Always = 2 Sometimes = 1 Never = 0	2
iv. Article II: Has the country, by its actions or the actions of those within its jurisdiction, allowed for the distribution on or from its territory programme-carrying signals by distributors for whom the emitted signal to or passing through the satellite is not intended?	Always = 0 Sometimes = 1 Never = 2	2
Articles 3, 4, 5, 6, and 7 describe the conditions in which Article 2 applies, therefore no scoring is assessed.		
Articles 8, 9, 10, 11 and 12: These are administrative articles, therefore no scoring is assessed.		
Treaty Score (Sum the above scores)		8
c. Institution-related treaties		
Individual treaties within this subsection are not scored as they are institution-specific and in some cases outdated and therefore irrelevant. However, collectively if a country ratified or acceded to at least two of these treaties, it will be assessed a score of 2 to demonstrate its participation and cooperation in international legal instruments and organisations related to space, a key principle promoted in international space law documents. Therefore, the following treaties for consideration are: i. Agreement Relating to the International Telecommunications Satellite Organization (ITSO) ii. Agreement on the Establishment of the INTERSPUTNIK International System and Organization of Space Communications iii. Convention for the Establishment of a European Space Agency (ESA) iv. Agreement of the Arab Corporation for Space Communications (ARABSAT) v. Agreement on Cooperation in the Exploration and Use of Outer Space for Peaceful Purposes (INTERCOSMOS) vi. Convention on the International Mobile Satellite Organization vii. Convention Establishing the European Telecommunications Satellite Organization (EUTELSAT) viii. Convention for the Establishment of a European Organization for the Exploitation of Meteorological Satellites (EUMETSAT) ix. International Telecommunication Constitution and Convention		0.6
Total (Sum the treaty scores from (a) and (b) and multiply by .70 and then sum with © score; bring total to Table B.1)		11.8

Appendix I: Scoring for United States

B.2	Non-Binding International Law Score
(a) Principles Governing the Use by States of Artificial Earth Satellites for International Direct Television Broadcasting *(Enter total from Table B.2.1 and multiply by .10)*	Advanced: 1.30 / 1.80
(b) Principles Relating to Remote Sensing of the Earth from Outer Space *(Enter total from Table B.2.2 and multiply by .10)*	Advanced: 1.20 / 1.80
(c) Principles Relevant to the Use of Nuclear Power Sources in Outer Space *(Enter total from Table B.2.3 and multiply by .10)*	Advanced: 3.0 / 3.0
(d) Declaration on International Cooperation in the Exploration and Use of Outer Space for the Benefit and in the Interest of All States, Taking into Particular Account the Needs of Developing Countries *(Enter total from Table B.2.4 and multiply by .10)*	Advanced: 0.5 / 1.0
(e) International cooperation in the peaceful uses of outer space: Some aspects concerning the use of the geostationary orbit *(Enter total from Table B.2.5 and multiply by .10)*	Advanced: 0.3 / 0.4
(f) Application of the concept of the "launching State" *(Enter total from Table B.2.6 and multiply by .10)*	Advanced: 0.5 / 0.6
(g) Recommendations on enhancing the practice of States and international intergovernmental organizations in registering space objects *(Enter total from Table B.2.7 and multiply by .10)*	Advanced: 0.8 / 0.8
(h) Recommendations on national legislation relevant to the peaceful exploration and use of outer space *(Enter total from Table B.2.8 and multiply by .10)*	Advanced: 1.30 / 1.40
(i) Space Debris Mitigation Guidelines of the Committee on the Peaceful Uses of Outer Space *(Enter total from Table B.2.9 and multiply by .10)*	Advanced: 1.40 / 1.40
(j) Safety Framework for Nuclear Power Source Applications in Outer Space *(Enter total from Table B.2.10 and multiply by .10)*	Advanced: 2.0 / 2.0
(k) Voluntary guidelines for transparency and confidence-building, space code of conduct, and long-term sustainability of space *(Enter total from Table B.2.11)*	Advanced: 2 / 6
Total *(Sum scores from above and bring total to Table A)*	Advanced: 14.3 / 20.4

B.2.1	Principles Governing the Use by States of Artificial Earth Satellites for International Direct Television Broadcasting	Score
1. Are the country's satellite direct television broadcasting activities respectful of state sovereignty, and do they follow the principle of non-intervention?	Always / not applicable = 2 Sometimes = 1 Never = 0	1
2. Do the country's satellite direct television broadcasting activities promote the free dissemination and exchange of information and knowledge in cultural and scientific fields?	Always / not applicable = 2 Sometimes = 1 Never = 0	1
3. Does the country's satellite direct television broadcasting activities assist in education and social and economic development?	Always / not applicable = 2 Sometimes = 1 Never = 0	1
4. Are the country's satellite direct television broadcasting activities carried out on a friendly and cooperative basis with other countries?	Always / not applicable = 2 Sometimes = 1 Never = 0	1
5. Does the country have in a place an authorisation system through its laws or administrative regulations or procedures for the conduct of satellite direct television activities?	Full implementation = 2 Partial implementation = 1 No implementation = 0	2
6. Does the country consult with other states regarding broadcasting or receiving of satellite direct television broadcasting for purposes of coordinating the same service?	Always / not applicable = 2 Sometimes = 1 Never = 0	2
7. Does the country participate in agreements with other states for the protection of copyright and similar rights?	Always / not applicable = 2 Sometimes = 1 Never = 0	2
8. Does the country notify the United Nations of satellite direct television broadcasting?	Always / not applicable = 2 Sometimes = 1 Never = 0	2
9. Does the country notify and consult the receiving state of a satellite direct television signal of its intent to broadcast?	Always / not applicable = 2 Sometimes = 1 Never = 0	1
Principles Score (Sum the above scores and bring to Table B.2)		13

B.2.2	Principles Relating to Remote Sensing of the Earth from Outer Space	Score
1. Does the country carry out remote sensing activities for the benefit and in the interests of all countries?	Always= 2 Sometimes = 1 Never = 0	1

Appendix I: Scoring for United States

2. Does the country's remote sensing activities observe the principle of full and permanent sovereignty of all states over their own wealth and natural resources?	Always = 2 Sometimes = 1 Never = 0		1
3. Does the country allow for international cooperation and participation in its remote sensing activities?	Always = 2 Sometimes = 1 Never = 0		1
4. Does the country enter into agreements with other states for the establishment and operation of data collection and storage stations and processing and interpretation facilities?	Always = 2 Sometimes = 1 Never = 0		1
5. Does the country offer technical assistance on a regular basis to other states interested in conducting remote sensing activities?	Always = 2 Sometimes = 1 No = 0		1
6. Does the country inform the United Nations of its remote sensing activities?	Always = 2 Sometimes = 1 Never = 0		1
7. Does the country disclose information to other states derived from remote sensing that could help avert harm to the Earth's natural environment?	Always = 2 Sometimes = 1 Never = 0		2
8. Does the country disclose information to other states derived from processing and analysing remote sensing data that is useful to states affected, or likely to be affected, by natural disasters?	Always = 2 Sometimes = 1 Never = 0		2
9. Does the country make available to remotely sensed states primary and processed data on the sensed state according to non-discriminatory and reasonable cost terms?	Always = 2 Sometimes = 1 Never = 0		1
10. Does the country allow for opportunities for a remotely sensed state to participate and benefit from activities that remotely sense it, including but not limited to access to remotely sensed data and participation in remote sensing activities ?	Always = 2 Sometimes = 1 Never = 0		1
Principles Score (Sum the above scores and bring to Table B.2)			12

B.2.3	Principles Relevant to the Use of Nuclear Power Sources in Outer Space		Score
1. Does the country design and use space objects with nuclear power sources with high degrees of confidence to avoid hazardous levels associated with operational or accidental circumstances?	Always / not applicable = 2 Sometimes = 1 Never = 0		2
2. Do the country's space objects with nuclear power sources follow design and construction that take into account generally accepted international radiological protection guidelines?	Always / not applicable = 2 Sometimes = 1 Never = 0		2
3. Do the country's space objects with nuclear power sources possess the capability of correcting or counteracting failures or malfunctions?	Always / not applicable = 2 Sometimes = 1 Never = 0		2
4. Do the country's space objects with nuclear power sources include systems that follow practices of redundancy, physical separation, functional isolation, and adequate independence of components?	Always / not applicable = 2 Sometimes = 1 Never = 0		2
5. Does the country only operate nuclear power sources on space objects that are on interplanetary missions, in sufficiently high orbits, and in low-Earth orbits that are later stored in sufficiently high orbits after their mission operations?	Always / not applicable= 2 Sometimes = 1 Never = 0		2
6. Does the country's space objects with nuclear power sources only use highly enriched uranium 235 as fuel?	Always / not applicable = 2 Sometimes = 1 Never = 0		2
7. Does the country only make critical its nuclear reactors once they have reached their operating orbit or interplanetary trajectory?	Always / not applicable= 2 Sometimes = 1 Never = 0		2
8. Does the country design and constructs its satellites with nuclear power sources to ensure that they do not become critical before reaching their operating orbit or interplanetary trajectory?	Always / not applicable = 2 Sometimes = 1 Never = 0		2
9. Does the country's space objects with nuclear reactors include an effective and controlled disposal of the nuclear reactor for those in an operational orbit less than sufficiently high?	Always / not applicable = 2 Sometimes = 1 Never = 0		2
10. Does the country only use radioisotope generators in space objects that are in interplanetary missions, or in missions in Earth orbit but which are later disposed of in high orbit?	Always / not applicable = 2 Sometimes = 1 Never = 0		2
11. Does the country's space object with radioisotope generators include containment systems that can withstand the heat and aerodynamic forces of re-entry, and which prevent the scattering of radioactive material in the environment?	Always / not applicable = 2 Sometimes = 1 Never = 0		2
12. Does the country conduct a comprehensive safety assessment for all phases of a mission for a space object with a nuclear power source?	Always / not applicable = 2 Sometimes = 1 Never = 0		2
13. Does the country indicate to the public and inform the United Nations of when it will launch a space object with a nuclear power source?	Always / not applicable = 2 Sometimes = 1 Never = 0		2
14. Does the country provide timely and updated information on space objects with nuclear power sources that malfunction and which may re-enter?	Always / not applicable = 2 Sometimes = 1 Never = 0		2
15. Does the country provide assistance to eliminate actual and possible harmful effects of re-entered space objects with nuclear power sources?	Always / not applicable = 2 Sometimes = 1 Never = 0		2
Principles Score (Sum the above scores and bring to Table B.2)			30

B.2.4	Declaration on International Cooperation in the Exploration and Use of Outer Space for the Benefit and in the Interest of All States, Taking into Particular Account the Needs of Developing Countries		Score
1. Does the country engage in equitable and mutually acceptable cooperation with other states according to fair and reasonable contractual terms?	Always = 2 Sometimes = 1 Never = 0		1
2. Does the country offer developing countries and countries with incipient space programmes opportunities to participate in advanced space activities?	Always = 2 Sometimes = 1 Never = 0		1
3. Do the country's space activities promote the development of space science and technology and its applications?	Always = 2 Sometimes = 1 Never = 0		1
4. Do the country's space activities help develop relevant and appropriate capabilities in states interested in space?	Always = 2 Sometimes = 1 Never = 0		1
5. Do the country's space activities facilitate the exchange of expertise and technology on a mutually acceptable basis?	Always = 2 Sometimes = 1 No = 0		1
Declaration Score (Sum the above scores and bring to Table B.2)			5

B.2.5	International cooperation in the peaceful uses of outer space: Some aspects concerning the use of the geostationary orbit		Score
1. Does the country already having access to a geostationary orbit/spectrum take all practicable steps to enable a developing country, which wishes to use the same, to have equitable access to the requested orbit/spectrum?	Always / not applicable = 2 Sometimes = 1 Never = 0		1
2. Does the country file satellite orbit and frequency requests according to International Telecommunications Union regulations?	Always / not applicable = 2 Sometimes = 1 Never = 0		2
Resolution Score (Sum the above scores and bring to Table B.2)			3

B.2.6	Application of the concept of the "launching State"		Score
1. Does the country have in place national laws authorizing and providing for continuing supervision of non-governmental entities operating in outer space?	Full implementation = 2 Partial implementation = 1 No implementation = 0		2
2. Does the country voluntarily provide information for the on-orbit transfer of ownership of space objects?	Always = 2 Sometimes = 1 Never = 0		2
3. Does the country have harmonized national space laws with international space laws?	Full harmonization = 2 Partial harmonization = 1 No harmonization = 0		1
Resolution Score (Sum the above scores and bring to Table B.2)			5

B.2.7	Recommendations on enhancing the practice of States and international intergovernmental organizations in registering space objects		Score
1. Does the country provide detailed information on its registered space objects to the United Nations, including: international designators, Coordinated Universal Time of launch date, kilometres and minutes and degrees of orbital parameters, geostationary orbit location (as applicable), change in status of operations, date decay or re-entry, date and condition of moving to a disposal orbit, and web links to space object information?	All information = 1 Partial information = 0 No information = -1		2
2. Does the country from which the territory or facility a space object is launched jointly determine with other states or intergovernmental organisations which entity should register the space object?	Always / not applicable = 2 Sometimes = 1 Never = 0		2
3. If participating in a joint launch, does the country register the space object separately from its joint launching state?	Always / not applicable = 2 Sometimes = 1 Never = 0		2
4. If a space object changes supervision, does the country of registry provide to the United Nations detailed information, including date of change in supervision, identification of the new owner or operator, change in orbital position, and change of function?	All information = 2 Partial information = 1 No information = 0		2
Resolution Score (Sum the above scores and bring to Table B.2)			8

B.2.8	Recommendations on national legislation relevant to the peaceful exploration and use of outer space		Score
1. Does the country have a national framework that governs the launch of objects into and their return from outer space, the operation of launch or re-entry sites and the operation and control of space objects in orbit, design and manufacture of spacecraft, application of space science and technology, and exploration activities and research?	All = 2 Some = 1 None = 0		2
2. Does the country issue authorisations and ensure supervision over space activities carried out from its jurisdiction or elsewhere when carried out by its citizens or legally established persons?	Always = 2 Sometimes = 1 Never = 0		2
3. Does the country have a national authority that authorizes space activities, including procedures for granting, modifying, suspending, and revoking authorisations?	All procedures = 2 Some procedures = 1 No procedures = 0		2

		Score
4. Does the country's authorisation regime ensure the safe conduct and minimal risk to persons for space activities, including assessing the expertise and technical qualifications of space operators and requiring adherence to safety and technical standards in line with the Space Debris Mitigation Guidelines?	Fully ensures = 2 Partially ensures = 1 Does not ensure = 0	1
5. Does the country's supervision and authorisation regime include on-site inspections or general reporting requirements?	Fully includes = 2 Partially includes = 1 Does not include = 0	2
6. Does the country's laws or administrative regulations or procedures offer recourse from operators or owners of space objects if the country's liability is engaged in the case of damages, such as through the use of insurance requirements or indemnification procedures?	Full recourse = 2 Partial recourse = 1 No recourse = 0	2
7. Does the country's supervision regime apply in case of transfer of ownership of space objects of non-governmental entities?	Always / not applicable = 2 Sometimes = 1 Never = 0	2
Recommendations Score (Sum the above scores and bring to Table B.2)		13

B.2.9	Space Debris Mitigation Guidelines of the Committee on the Peaceful Uses of Outer Space	Score
1. Do the country's laws or administrative regulations or procedures require space systems to be designed to not release debris during normal operations?	Always = 2 Sometimes = 1 Never = 0	2
2. Do the country's laws or administrative regulations or procedures require spacecraft and launch vehicle orbital stages to be designed to avoid accidental break-ups?	Always = 2 Sometimes = 1 Never = 0	2
3. Do the country's laws or administrative regulations or procedures require spacecraft and launch vehicle orbital stages to have disposal and passivation measures in the case of failure?	Always = 2 Sometimes = 1 Never = 0	2
4. Do the country's laws or administrative regulations or procedures require spacecraft operators to follow collision avoidance procedures through adjustment of launch times and on-orbit avoidance manoeuvres?	Always = 2 Sometimes = 1 Never = 0	2
5. Does the country's laws or administrative regulations or procedures require space objects to deplete or make safe their on-board sources of stored energy, including through passivation by removal of residual propellants, compressed fluids, and electrical storage devices?	Always = 2 Sometimes = 1 Never = 0	2
6. Do the country's laws or administrative regulations or procedures limit the long-term presence of space objects in low-Earth orbit after their end of mission, including through controlled removal from orbit through re-entry or in orbits beyond low-Earth orbit?	Always = 2 Sometimes = 1 Never = 0	2
7. Do the country's laws or administrative regulations or procedures limit the long-term presence of space objects in geosynchronous Earth orbit after their end of mission, including through controlled removal to orbits beyond geosynchronous Earth orbit?	Always = 2 Sometimes = 1 Never = 0	2
Guidelines Score (Sum the above scores and bring to Table B.2)		14

B.2.10	Safety Framework for Nuclear Power Source Applications in Outer Space	Score
1. Does the country have in place safety policies, requirements, and processes that protect people and the environment in Earth's biosphere from potential hazards associated with relevant launch, operation, and end-of-service phases of space nuclear power source applications?	All phases / not applicable = 2 Some phases = 1 No phases = 0	2
2. Does the country's space activities authorisation regime verify the rationale and require justification for the use of space nuclear power source applications?	Always / not applicable = 2 Sometimes = 1 Never = 0	2
3. Does the country's space launch authorisation regime require an independent safety evaluation assessing the risk to people and the environment for launches, operations, and end-of-service phases for space nuclear power source applications?	Always / not applicable = 2 Sometimes = 1 Never = 0	2
4. Does the country conduct emergency preparedness activities, including emergency planning, training, rehearsals and development of procedures and communication protocols, in preparation for radiation exposure to people and the environment as a result of space nuclear power applications?	Repeatedly = 2 Once = 1 Never = 0	2
5. Does the country's laws or administrative regulations or procedures identify space operators of space nuclear power source applications as the primary responsibility holders for operations of such applications?	Full recognition = 2 Partial recognition = 1 No recognition = 0	2
6. Does the country's laws or administrative regulations or procedures require space operators to include safety management to form part of overall space mission management?	Always / not applicable = 2 Sometimes = 1 Never = 0	2
7. Does the country's laws or administrative regulations or procedures require space operations to have technical competence in nuclear safety, including qualified individuals and facilities for designing, testing, and analysing nuclear safety capabilities part of space missions?	Always / not applicable = 2 Sometimes = 1 Never = 0	2
8. Does the country's laws or administrative regulations or procedures require space operators to integrate safety considerations from design to development to launch and operations and end-of-service for the entire space nuclear power source application?	Always / not applicable = 2 Sometimes = 1 Never = 0	2
9. Does the country's laws or administrative regulations or procedures require space operators to conduct risk assessments on the launch, operation, and end-of-service phases of space nuclear power source applications?	Always / not applicable = 2 Sometimes = 1 Never = 0	2
10. Does the country's laws or administrative regulations or procedures require mitigation measures for accidents of space nuclear power source applications with the potential to release radioactive material into the environment?	Always / not applicable = 2 Sometimes = 1 Never = 0	2
Framework Score (Sum the above scores and bring to Table B.2)		20

B.2.11	Voluntary guidelines for transparency and confidence-building, space code of conduct, and long-term sustainability of space	Score
Individual recommendations included within these three separate sets of guidelines are not scored. However, if a country has implemented a majority of the individual recommendations contained within each set of guidelines in either its legal or administrative practice, it will be assessed a score of 2 per set of guidelines. Therefore, the following sets of guidelines for consideration are:		

Appendix I: Scoring for United States

i. Recommendations from the Group of Governmental Experts on Transparency and Confidence-Building Measures	0
ii. Recommendations from the European Union's International Code of Conduct on Outer Space Activities	0
iii. Recommendations from the UNCOPUOS Long-Term Sustainability Guidelines	2
Guidelines Score (Sum the above scores and bring to Table B.2)	2

Appendix II: Scoring for South Korea

A	Aggregate Score
Binding International Law Score *(Enter total from last line of Table B.1)*	Intermediate: 18.28 / 18.90
Non-Binding International Law Score *(Enter total from last line of Table B.2)*	Intermediate: 12.4 / 20.40
Total *(Sum binding and non-binding international law scores)*	Intermediate: 30.68 / 39.30

Appendix II: Scoring for South Korea

B.1	Binding International Law Score
Space-specific International Treaties *(Use total from last line of Table B.1.1 and multiply by .50)*	Intermediate: 12.38 / 13.0
Space-related International Treaties *(Use total from last line of Table B.1.2 and multiply by .50)*	Intermediate: 5.90 / 5.90
Total *(Sum space-specific and space-related treaties scores and then bring total to Table A)*	Intermediate: 18.28 / 18.90

B.1.1	Space-specific International Treaties	Score
1. Treaty on Principles Governing the Activities of States in the Exploration and Use of Outer Space, including the Moon and Other Celestial Bodies (25 percent of section score)		
a. Has the country ratified or acceded to this treaty?	Yes, fully = 2 Yes, with reservations = 1 No = 0	2
b. Is the treaty domesticated in national laws or administrative regulations or procedures?	Full domestication = 2 Partial domestication = 1 No domestication = 0	2
c. Article I: Do country's laws or administrative regulations or procedures recognise the role of international law?	Full recognition = 2 Partial recognition = 1 No recognition = 0	2
d. Article I: Do the country's laws or administrative regulations or procedures affirm that the exploration and use of outer space should be carried out for the benefit and interests of all countries?	Full affirmation = 2 Partial affirmation = 1 No affirmation = 0	1
e. Article I: Do the country's laws or administrative regulations or procedures recognise outer space as the province of all mankind?	Full recognition = 2 Partial recognition = 1 No affirmation = 0	0
f. Article I: Do the country's laws or administrative regulations or procedures recognise that outer space should be free for exploration without discrimination and with free access to celestial bodies?	Full recognition = 2 Partial recognition = 1 No recognition = 0	2
g. Article I: Do the country's laws or administrative regulations or procedures recognise the freedom of scientific investigation in outer space?	Full recognition = 2 Partial recognition = 1 No recognition = 0	2
h. Article I: Does the country facilitate and participate in international scientific investigation?	Frequently = 2 Occassionally = 1 Never = 0	2
i. Article II: Do the country's laws or administrative regulations or procedures allow for, or encourage, the appropriation of outer space and celestial bodies by claims of sovereignty, use, or occupation?	Always = 0 Sometimes = 1 Never = 2	
j. Article II: Does the country, by its actions or the actions of those within its jurisdiction, appropriate, or encourage the appropriation of, outer space and celestial bodies and resources by claims of sovereignty, use, or occupation?	Frequently = 0 Occassionally = 1 No = 2	
Article III: Elements of this article are already captured in questions asked above, therefore no separate scoring is assessed.	N/A	
k. Article IV: Has the country placed into orbit or installed on celestial bodies weapons of mass destruction, including nuclear weapons?	Repeatedly = 0 Once = 1 No = 2	2
l. Article IV: Has the country established bases, installations, or fortifications, or tested weapons or conducted military manoeuvres on celestial bodies?	Repeatedly = 0 Once = 1 No = 2	
m. Article IV: Do the country's laws or administrative regulations or procedures allow for the conduct of such types of activities as described in Question 'l' above?	Always = 0 Sometimes = 1 Never = 2	
n. Article V: Has the country either provided assistance or rejected assistance to astronauts of other countries who are in distress?	Always provided / not applicable = 2 Sometimes provided = 1 Never provided = 0	2
o. Article V: Does the country proactively inform the United Nations and other countries regarding potentially dangerous phenomena to astronauts?	Always = 2 Sometimes = 1 Never = 0	2
p. Article VI: Does the country have in place an authorisation and supervision regime for space activities conducted by non-governmental entities?	Yes = 2 Partial = 1 No = 0	2
q. Article VII: Does the country recognise its liability as a launching state in its laws or administrative regulations or procedures?	Full recognition = 2 Partial recognition = 1 No recognition = 0	2
r. Article VII: Has the country denied its liability as a launching state for damage caused?	Always = 0 Sometimes = 1 Never = 2	2
s. Article VIII: Does the country maintain an updated national registry of launched space objects?	Always = 2 Sometimes = 1 Never = 0	2
t. Article VIII: Has the country ever returned a space object belonging to another country, but which landed in its jurisdiction?	Always / not applicable = 2 Sometimes = 1 Never = 0	2
u. Article IX: Do the country's laws or administrative regulations or procedures account for avoiding harmful contamination, including debris, of the space environment?	Full account = 2 Partial account = 1 No account = 0	1
v. Article IX: Do the country's laws or administrative regulations or procedures seek to limit damage to the Earth environment by extraterrestrial material?	Full account = 2 Partial account = 1 No account = 0	2
w. Article X: Does the country allow other countries to observe the flight of space objects that it launches?	Always = 2 Sometimes = 1 Never = 0	1
x. Article XI: Does the country inform the United Nations and other countries of its space activities, including the nature, conduct, location, and results of the activities?	Always = 2 Sometimes = 1 Never = 0	1
y. Article XII: Does the country allow access and visits to its outer space stations, installations, equipment, and vehicles, provided reasonable notification of a such visit is given by the visiting representatives?	Always / not applicable = 2 Sometimes = 1 Never = 0	2
Articles XIII, XIV, XV, XVI, and XVII: These are administrative articles, therefore no scoring is assessed.		
Treaty Score (Sum the above scores and multiply by .25)		9
2. Agreement on the Rescue of Astronauts, the Return of Astronauts and the Return of Objects Launched into Outer Space *(25 percent of section score)*		
	Yes, in full = 2	

Appendix II: Scoring for South Korea

a. Has the country ratified or acceded to this treaty?	Yes, with reservations = 1 No = 0		2
b. Is the treaty domesticated in national laws or administrative regulations or procedures?	Full domestication = 2 Partial domestication = 1 No domestication = 0		2
c. Article 1: Has the country notified the launching state(s) of astronauts who have suffered accidents, been in distress, or made emergency landings?	Always / not applicable = 2 Sometimes = 1 Never = 0		2
d. Article 1: Has the country notified the United Nations of astronauts who have suffered accidents, been in distress, or made emergency landings?	Always / not applicable = 2 Sometimes = 1 No = 0		2
e. Article 2: Has the country taken all possible steps to rescue and render all necessary assistance to astronauts who have suffered accidents, been in distress, or made emergency landings?	Always / not applicable = 2 Sometimes = 1 No = 0		2
f. Article 3: Has the country provided assistance in search and rescue operations for astronauts who have suffered accidents, been in distress, or made emergency landings?	Always / not applicable = 2 Sometimes = 1 No = 0		2
g. Article 4: Has the country safely and promptly returned to their launching states astronauts who have suffered accidents, been in distress, or made emergency landings?	Always / not applicable = 2 Sometimes = 1 No = 0		2
h. Article 5: Has the country notified the launching state and United Nations of a space object or part of a space object that has returned to Earth?	Always / not applicable = 2 Sometimes = 1 No = 0		2
i. Article 5: Has the country in which the returning space object landed practicably assisted to recover the object or its parts?	Always / not applicable = 2 Sometimes = 1 No = 0		2
j. Article 5: Has the country in which the returning space object landed returned the space object or its parts and provided their identifying data to the launching state?	Always / not applicable = 2 Sometimes = 1 No = 0		2
k. Article 5: Has the country in which the returning space object landed eliminated the possibility of its danger if it is found to be hazardous or deleterious?	Always / not applicable = 2 Sometimes = 1 No = 0		2
l. Article 5: Has the launching state paid for expenses related to the recovery and return of space object or its part incurred by the country in which the space object landed?	Always / not applicable = 2 Sometimes = 1 No = 0		2
Articles 6, 7, 8, 9, and 10: These are administrative articles, therefore no scoring is assessed.			
Treaty Score (Sum the above scores and multiply by .25)			6
3. Convention on International Liability for Damage Caused by Space Objects			
a. Has the country ratified or acceded to this treaty?	Yes, in full = 2 Yes, with reservations = 1 No = 0		2
b. Is the treaty domesticated in national laws or administrative regulations or procedures?	Full domestication = 2 Partial domestication = 1 No domestication = 0		2
Article I provides for definitions of terms used in the treaty, therefore no scoring is assessed.			
c. Article II: Do the country's laws or administrative regulations or procedures recognise its liability to pay compensation for damage caused by its space objects on the Earth's surface or aircraft in flight?	Full recognition = 2 Partial recognition = 1 No recognition = 0		2
d. Article II: Has the country refused to pay compensation in practice for damage caused by its space objects on the Earth's surface or aircraft in flight?	Always = 0 Sometimes = 1 Never / not applicable = 2		2
e. Article III: Do the country's laws or administrative regulations or procedures recognise its liability owing to its fault or fault of persons for which it is responsible in the case of damage to another space object, persons, or property in space of another launching state?	Full recognition = 2 Partial recognition = 1 No recognition = 0		2
f. Article III: Has the country refused to accept liability in the case of damage to another space object, persons, or property in space of another launching state caused by its fault or fault of persons for which it is responsible?	Always = 0 Sometimes = 1 Never / not applicable = 2		2
g. Article IV: Do the country's laws or administrative regulations or procedures recognise its absolute joint and several liability with a second-party launching state if its space object caused damage to the second-party launching state's space object, which then resulted in damage on the Earth's surface or aircraft in flight of a third state?	Full recognition = 2 Partial recognition = 1 No recognition = 0		2
h. Article IV: Does the country's laws or administrative regulations or procedures recognise its joint and several liability apportioned according to fault with a second-party launching state if its space object caused damage to the second-party launching state's space object, which then resulted in damage to a third state's space object, person, or property in space?	Full recognition = 2 Partial recognition = 1 No recognition = 0		2
i. Article IV: Has the country refused to pay compensation according to its fault in cases of joint and several liability for damage caused to a third-party state?	Always = 0 Sometimes = 1 Never / not applicable = 2		2
j. Article V: Do the country's laws or administrative regulations or procedures recognise joint and several liability in cases when it and another state jointly launch a space object?	Full recognition = 2 Partial recognition = 1 No recognition = 0		2
k. Article V: Do country's laws or administrative regulations or procedures account for its roles a joint launch participant if a space object is launched from its territory or facility?	Full account = 2 Partial account = 1 No account = 0		2
Articles VI and VII sets forth conditions under which other articles apply and is not actionable by states, therefore no scoring is assessed.			
l. Article VIII: Do the country's laws or administrative regulations or procedures provide for a claims and compensation process for damage which it may suffer by other launching states?	Full provision = 2 Partial provision = 1 No provision = 0		2
Article IX, X, XI, XII, and XIII prescribe the administrative details through which claims and compensation should be processed, therefore no scoring is assessed.			
Articles XIV, XV, XVI, XVII, and XVIII prescribe the administrative details through which a Claims Commission should be established and make adjudications in the case of no settlements through diplomatic channels established elsewhere in the treaty, therefore no scoring is assessed.			
m. Article XIX: Has the country refused or failed to pay compensation or an award decided by a Claims Commission as set forth in other treaty articles?	Always = 0 Sometimes = 1 Never / not applicable = 2		2
Article XX prescribes the process through which the expenses associated with the Claims Commission should be borne by state participants, therefore no scoring is assessed.			
n. Article XXI: Has the country rendered appropriate and rapid assistance in cases when a space object presents large-scale danger to human life or interferes with a population's living conditions or functioning of a vital centre?	Always / not applicable = 2 Sometimes = 1 Never = 0		2
Articles XXII, XXIII, XXIV, XXV, XXVI, XXVII, XXVIII: These are administrative articles, therefore no scoring is assessed.			
Treaty Score (Sum the above scores and multiply by .25)			7
4. Convention on Registration of Objects Launched into Outer Space			
a. Has the country ratified or acceded to this treaty?	Yes, in full = 2 Yes, with reservations = 1 No = 0		2

Appendix II: Scoring for South Korea

	Options		Score
b. Is the treaty domesticated in national laws or administrative regulations or procedures?	Full domestication = 2 Partial domestication = 1 No domestication = 0		2
Article I provides for definitions of terms used in the treaty, therefore no scoring is assessed.			
c. Article II: Does the country maintain an updated national registry of launched space objects that it launches and that it launches jointly with other launching states?	Always / not applicable = 2 Sometimes = 1 Never = 0		2
Article III describes actions to be taken by the United Nations rather than a specific country, therefore no scoring is assessed.			
d. Article IV: Does the country provide detailed information to the United Nations for its launched space objects, including name of the country, space object registration number, date and territory of launch, basic orbital parameters, and general function of the space object?	Full information / not applicable = 2 Partial information = 1 No information = 0		1
e. Article IV: Does the country update the United Nations on space objects for which it previously shared information, but which are no longer in Earth orbit?	Always / not applicable = 2 Sometimes = 1 Never = 0		2
Article V describes a procedural administrative action, therefore no scoring is assessed.			
f. Article VI: Does the country assist with providing identifying information it has gained through space monitoring and tracking facilities for space objects that are unidentifiable and which are of a hazardous or deleterious nature?	Always = 2 Sometimes = 1 Never = 0		2
Articles VII, VIII, IX, X, and XII: These are administrative articles, therefore no scoring is assessed.			
Treaty Score (Sum the above scores and multiply by .25)			2.75
Total (Sum the above treaty score and bring total to Table B.1)			24.75

B.1.2	Space-related International Treaties		Score
a. Treaty Banning Nuclear Weapon Tests in the Atmosphere, in Outer Space and under Water			
i. Has the country ratified or acceded to this treaty?	Yes, in full = 2 Yes, with reservations = 1 No = 0		2
ii. Is the treaty domesticated in national laws or administrative regulations or procedures?	Full domestication = 2 Partial domestication = 1 No domestication = 0		2
iii. Article I: Do the country's laws or administrative regulations or procedures prohibit, prevent, and forbid the carrying out of nuclear weapon tests at places under its jurisdiction or control, including in the atmosphere and outer space and underwater?	Full prohibition = 2 Partial prohibition = 1 No prohibition = 0		2
iv. Article I: Since implementing the treaty, has the country carried out a nuclear weapon test at a place under its jurisdiction or control, including in the atmosphere and outer space and underwater?	Repeatedly = 0 Once = 1 Never = 2		2
Articles II, III, IV, and V: These are administrative articles, therefore no scoring is assessed.			
Treaty Score (Sum the above scores)			8
b. Convention Relating to the Distribution of Programme-Carrying Signals Transmitted by Satellite			
i. Has the country ratified or acceded to this treaty?	Yes, in full = 2 Yes, with reservations = 1 No = 0		2
ii. Is the treaty domesticated in national laws or administrative regulations or procedures?	Full domestication = 2 Partial domestication = 1 No domestication = 0		2
Article 1 provides for definitions of terms used in the treaty, therefore no scoring is assessed.			
iii. Article II: Do the country's laws or administrative regulations or procedures prevent the distribution on or from its territory programme-carrying signals by distributors for whom the emitted signal to or passing through the satellite is not intended?	Always = 2 Sometimes = 1 Never = 0		2
iv. Article II: Has the country, by its actions or the actions of those within its jurisdiction, allowed for the distribution on or from its territory programme-carrying signals by distributors for whom the emitted signal to or passing through the satellite is not intended?	Always = 0 Sometimes = 1 Never = 2		2
Articles 3, 4, 5, 6, and 7 describe the conditions in which Article 2 applies, therefore no scoring is assessed.			
Articles 8, 9, 10, 11 and 12: These are administrative articles, therefore no scoring is assessed.			
Treaty Score (Sum the above scores)			8
c. Institution-related treaties			
Individual treaties within this subsection are not scored as they are institution-specific and in some cases outdated and therefore irrelevant. However, collectively if a country ratified or acceded to at least two of these treaties, it will be assessed a score of 2 to demonstrate its participation and cooperation in international legal instruments and organisations related to space, a key principle promoted in international space law documents. Therefore, the following treaties for consideration are:			
i. Agreement Relating to the International Telecommunications Satellite Organization (ITSO)			
ii. Agreement on the Establishment of the INTERSPUTNIK International System and Organization of Space Communications			
iii. Convention for the Establishment of a European Space Agency (ESA)			0.6
iv. Agreement of the Arab Corporation for Space Communications (ARABSAT)			
v. Agreement on Cooperation in the Exploration and Use of Outer Space for Peaceful Purposes (INTERCOSMOS)			
vi. Convention on the International Mobile Satellite Organization			
vii. Convention Establishing the European Telecommunications Satellite Organization (EUTELSAT)			
viii. Convention for the Establishment of a European Organization for the Exploitation of Meteorological Satellites (EUMETSAT)			
ix. International Telecommunication Constitution and Convention			
Total (Sum the treaty scores from (a) and (b) and multiply by .70 and then sum with © score; bring total to Table B.1)			11.8

Appendix II: Scoring for South Korea

B.2	Non-Binding International Law Score
(a) Principles Governing the Use by States of Artificial Earth Satellites for International Direct Television Broadcasting *(Enter total from Table B.2.1 and multiply by .10)*	Intermediate: 1.50 / 1.80
(b) Principles Relating to Remote Sensing of the Earth from Outer Space *(Enter total from Table B.2.2 and multiply by .10)*	Intermediate: 1.20 / 1.80
(c) Principles Relevant to the Use of Nuclear Power Sources in Outer Space *(Enter total from Table B.2.3 and multiply by .10)*	Intermediate: 3.0 / 3.0
(d) Declaration on International Cooperation in the Exploration and Use of Outer Space for the Benefit and in the Interest of All States, Taking into Particular Account the Needs of Developing Countries *(Enter total from Table B.2.4 and multiply by .10)*	Intermediate: 1.0 / 1.0
(e) International cooperation in the peaceful uses of outer space: Some aspects concerning the use of the geostationary orbit *(Enter total from Table B.2.5 and multiply by .10)*	Intermediate: 0.3 / 0.4
(f) Application of the concept of the "launching State" *(Enter total from Table B.2.6 and multiply by .10)*	Intermediate: 0.6 / 0.6
(g) Recommendations on enhancing the practice of States and international intergovernmental organizations in registering space objects *(Enter total from Table B.2.7 and multiply by .10)*	Intermediate: 0.8 / 0.8
(h) Recommendations on national legislation relevant to the peaceful exploration and use of outer space *(Enter total from Table B.2.8 and multiply by .10)*	Intermediate: 1.30 / 1.40
(i) Space Debris Mitigation Guidelines of the Committee on the Peaceful Uses of Outer Space *(Enter total from Table B.2.9 and multiply by .10)*	Intermediate: 0.7 / 1.40
(j) Safety Framework for Nuclear Power Source Applications in Outer Space *(Enter total from Table B.2.10 and multiply by .10)*	Intermediate: 2.0 / 2.0
(k) Voluntary guidelines for transparency and confidence-building, space code of conduct, and long-term sustainability of space *(Enter total from Table B.2.11)*	Intermediate: 0 / 6
Total *(Sum scores from above and bring total to Table A)*	Intermediate: 12.4 / 20.4

B.2.1	Principles Governing the Use by States of Artificial Earth Satellites for International Direct Television Broadcasting	Score
1. Are the country's satellite direct television broadcasting activities respectful of state sovereignty, and do they follow the principle of non-intervention?	Always / not applicable = 2 Sometimes = 1 Never = 0	1
2. Do the country's satellite direct television broadcasting activities promote the free dissemination and exchange of information and knowledge in cultural and scientific fields?	Always / not applicable = 2 Sometimes = 1 Never = 0	2
3. Does the country's satellite direct television broadcasting activities assist in education and social and economic development?	Always / not applicable = 2 Sometimes = 1 Never = 0	2
4. Are the country's satellite direct television broadcasting activities carried out on a friendly and cooperative basis with other countries?	Always / not applicable = 2 Sometimes = 1 Never = 0	0
5. Does the country have in a place an authorisation system through its laws or administrative regulations or procedures for the conduct of satellite direct television activities?	Full implementation = 2 Partial implementation = 1 No implementation = 0	2
6. Does the country consult with other states regarding broadcasting or receiving of satellite direct television broadcasting for purposes of coordinating the same service?	Always / not applicable = 2 Sometimes = 1 Never = 0	2
7. Does the country participate in agreements with other states for the protection of copyright and similar rights?	Always / not applicable = 2 Sometimes = 1 Never = 0	2
8. Does the country notify the United Nations of satellite direct television broadcasting?	Always / not applicable = 2 Sometimes = 1 Never = 0	2
9. Does the country notify and consult the receiving state of a satellite direct television signal of its intent to broadcast?	Always / not applicable = 2 Sometimes = 1 Never = 0	2
Principles Score (Sum the above scores and bring to Table B.2)		15

B.2.2	Principles Relating to Remote Sensing of the Earth from Outer Space	Score
1. Does the country carry out remote sensing activities for the benefit and in the interests of all countries?	Always= 2 Sometimes = 1 Never = 0	1

			Score
2. Does the country's remote sensing activities observe the principle of full and permanent sovereignty of all states over their own wealth and natural resources?	Always = 2 Sometimes = 1 Never = 0		1
3. Does the country allow for international cooperation and participation in its remote sensing activities?	Always = 2 Sometimes = 1 Never = 0		1
4. Does the country enter into agreements with other states for the establishment and operation of data collection and storage stations and processing and interpretation facilities?	Always = 2 Sometimes = 1 Never = 0		1
5. Does the country offer technical assistance on a regular basis to other states interested in conducting remote sensing activities?	Always = 2 Sometimes = 1 No = 0		1
6. Does the country inform the United Nations of its remote sensing activities?	Always = 2 Sometimes = 1 Never = 0		1
7. Does the country disclose information to other states derived from remote sensing that could help avert harm to the Earth's natural environment?	Always = 2 Sometimes = 1 Never = 0		2
8. Does the country disclose information to other states derived from processing and analysing remote sensing data that is useful to states affected, or likely to be affected, by natural disasters?	Always = 2 Sometimes = 1 Never = 0		2
9. Does the country make available to remotely sensed states primary and processed data on the sensed state according to non-discriminatory and reasonable cost terms?	Always = 2 Sometimes = 1 Never = 0		1
10. Does the country allow for opportunities for a remotely sensed state to participate and benefit from activities that remotely sense it, including but not limited to access to remotely sensed data and participation in remote sensing activities ?	Always = 2 Sometimes = 1 Never = 0		1
Principles Score (Sum the above scores and bring to Table B.2)			12

B.2.3	Principles Relevant to the Use of Nuclear Power Sources in Outer Space		Score
1. Does the country design and use space objects with nuclear power sources with high degrees of confidence to avoid hazardous levels associated with operational or accidental circumstances?	Always / not applicable = 2 Sometimes = 1 Never = 0		2
2. Do the country's space objects with nuclear power sources follow design and construction that take into account generally accepted international radiological protection guidelines?	Always / not applicable = 2 Sometimes = 1 Never = 0		2
3. Do the country's space objects with nuclear power sources possess the capability of correcting or counteracting failures or malfunctions?	Always / not applicable = 2 Sometimes = 1 Never = 0		2
4. Do the country's space objects with nuclear power sources include systems that follow practices of redundancy, physical separation, functional isolation, and adequate independence of components?	Always / not applicable = 2 Sometimes = 1 Never = 0		2
5. Does the country only operate nuclear power sources on space objects that are on interplanetary missions, in sufficiently high orbits, and in low-Earth orbits that are later stored in sufficiently high orbits after their mission operations?	Always / not applicable= 2 Sometimes = 1 Never = 0		2
6. Does the country's space objects with nuclear power sources only use highly enriched uranium 235 as fuel?	Always / not applicable = 2 Sometimes = 1 Never = 0		2
7. Does the country only make critical its nuclear reactors once they have reached their operating orbit or interplanetary trajectory?	Always / not applicable= 2 Sometimes = 1 Never = 0		2
8. Does the country design and constructs its satellites with nuclear power sources to ensure that they do not become critical before reaching their operating orbit or interplanetary trajectory?	Always / not applicable = 2 Sometimes = 1 Never = 0		2
9. Does the country's space objects with nuclear reactors include an effective and controlled disposal of the nuclear reactor for those in an operational orbit less than sufficiently high?	Always / not applicable = 2 Sometimes = 1 Never = 0		2
10. Does the country only use radioisotope generators in space objects that are in interplanetary missions, or in missions in Earth orbit but which are later disposed of in high orbit?	Always / not applicable = 2 Sometimes = 1 Never = 0		2
11. Does the country's space object with radioisotope generators include containment systems that can withstand the heat and aerodynamic forces of re-entry, and which prevent the scattering of radioactive material in the environment?	Always / not applicable = 2 Sometimes = 1 Never = 0		2
12. Does the country conduct a comprehensive safety assessment for all phases of a mission for a space object with a nuclear power source?	Always / not applicable = 2 Sometimes = 1 Never = 0		2
13. Does the country indicate to the public and inform the United Nations of when it will launch a space object with a nuclear power source?	Always / not applicable = 2 Sometimes = 1 Never = 0		2
14. Does the country provide timely and updated information on space objects with nuclear power sources that malfunction and which may re-enter?	Always / not applicable = 2 Sometimes = 1 Never = 0		2
15. Does the country provide assistance to eliminate actual and possible harmful effects of re-entered space objects with nuclear power sources?	Always / not applicable = 2 Sometimes = 1 Never = 0		2
Principles Score (Sum the above scores and bring to Table B.2)			30

Appendix II: Scoring for South Korea

B.2.4	Declaration on International Cooperation in the Exploration and Use of Outer Space for the Benefit and in the Interest of All States, Taking into Particular Account the Needs of Developing Countries		Score
1. Does the country engage in equitable and mutually acceptable cooperation with other states according to fair and reasonable contractual terms?	Always = 2 Sometimes = 1 Never = 0		2
2. Does the country offer developing countries and countries with incipient space programmes opportunities to participate in advanced space activities?	Always = 2 Sometimes = 1 Never = 0		2
3. Do the country's space activities promote the development of space science and technology and its applications?	Always = 2 Sometimes = 1 Never = 0		2
4. Do the country's space activities help develop relevant and appropriate capabilities in states interested in space?	Always = 2 Sometimes = 1 Never = 0		2
5. Do the country's space activities facilitate the exchange of expertise and technology on a mutually acceptable basis?	Always = 2 Sometimes = 1 No = 0		2
Declaration Score (Sum the above scores and bring to Table B.2)			10

B.2.5	International cooperation in the peaceful uses of outer space: Some aspects concerning the use of the geostationary orbit		Score
1. Does the country already having access to a geostationary orbit/spectrum take all practicable steps to enable a developing country, which wishes to use the same, to have equitable access to the requested orbit/spectrum?	Always / not applicable = 2 Sometimes = 1 Never = 0		1
2. Does the country file satellite orbit and frequency requests according to International Telecommunications Union regulations?	Always / not applicable = 2 Sometimes = 1 Never = 0		2
Resolution Score (Sum the above scores and bring to Table B.2)			3

B.2.6	Application of the concept of the "launching State"		Score
1. Does the country have in place national laws authorizing and providing for continuing supervision of non-governmental entities operating in outer space?	Full implementation = 2 Partial implementation = 1 No implementation = 0		2
2. Does the country voluntarily provide information for the on-orbit transfer of ownership of space objects?	Always = 2 Sometimes = 1 Never = 0		2
3. Does the country have harmonized national space laws with international space laws?	Full harmonization = 2 Partial harmonization = 1 No harmonization = 0		2
Resolution Score (Sum the above scores and bring to Table B.2)			6

B.2.7	Recommendations on enhancing the practice of States and international intergovernmental organizations in registering space objects		Score
1. Does the country provide detailed information on its registered space objects to the United Nations, including: international designators, Coordinated Universal Time of launch date, kilometres and minutes and degrees of orbital parameters, geostationary orbit location (as applicable), change in status of operations, date decay or re-entry, date and condition of moving to a disposal orbit, and web links to space object information?	All information = 1 Partial information = 0 No information = -1		2
2. Does the country from which the territory or facility a space object is launched jointly determine with other states or intergovernmental organisations which entity should register the space object?	Always / not applicable = 2 Sometimes = 1 Never = 0		2
3. If participating in a joint launch, does the country register the space object separately from its joint launching state?	Always / not applicable = 2 Sometimes = 1 Never = 0		2
4. If a space object changes supervision, does the country of registry provide to the United Nations detailed information, including date of change in supervision, identification of the new owner or operator, change in orbital position, and change of function?	All information = 2 Partial information = 1 No information = 0		2
Resolution Score (Sum the above scores and bring to Table B.2)			8

B.2.8	Recommendations on national legislation relevant to the peaceful exploration and use of outer space		Score
1. Does the country have a national framework that governs the launch of objects into and their return from outer space, the operation of launch or re-entry sites and the operation and control of space objects in orbit, design and manufacture of spacecraft, application of space science and technology, and exploration activities and research?	All = 2 Some = 1 None = 0		2
2. Does the country issue authorisations and ensure supervision over space activities carried out from its jurisdiction or elsewhere when carried out by its citizens or legally established persons?	Always = 2 Sometimes = 1 Never = 0		2
3. Does the country have a national authority that authorizes space activities, including procedures for granting, modifying, suspending, and revoking authorisations?	All procedures = 2 Some procedures = 1 No procedures = 0		2

4. Does the country's authorisation regime ensure the safe conduct and minimal risk to persons for space activities, including assessing the expertise and technical qualifications of space operators and requiring adherence to safety and technical standards in line with the Space Debris Mitigation Guidelines?	Fully ensures = 2 Partially ensures = 1 Does not ensure = 0	1
5. Does the country's supervision and authorisation regime include on-site inspections or general reporting requirements?	Fully includes = 2 Partially includes = 1 Does not include = 0	2
6. Does the country's laws or administrative regulations or procedures offer recourse from operators or owners of space objects if the country's liability is engaged in the case of damages, such as through the use of insurance requirements or indemnification procedures?	Full recourse = 2 Partial recourse = 1 No recourse = 0	2
7. Does the country's supervision regime apply in case of transfer of ownership of space objects of non-governmental entities?	Always / not applicable = 2 Sometimes = 1 Never = 0	2
Recommendations Score (Sum the above scores and bring to Table B.2)		13

B.2.9	Space Debris Mitigation Guidelines of the Committee on the Peaceful Uses of Outer Space	Score
1. Do the country's laws or administrative regulations or procedures require space systems to be designed to not release debris during normal operations?	Always = 2 Sometimes = 1 Never = 0	1
2. Do the country's laws or administrative regulations or procedures require spacecraft and launch vehicle orbital stages to be designed to avoid accidental break-ups?	Always = 2 Sometimes = 1 Never = 0	1
3. Do the country's laws or administrative regulations or procedures require spacecraft and launch vehicle orbital stages to have disposal and passivation measures in the case of failure?	Always = 2 Sometimes = 1 Never = 0	1
4. Do the country's laws or administrative regulations or procedures require spacecraft operators to follow collision avoidance procedures through adjustment of launch times and on-orbit avoidance manoeuvres?	Always = 2 Sometimes = 1 Never = 0	1
5. Does the country's laws or administrative regulations or procedures require space objects to deplete or make safe their on-board sources of stored energy, including through passivation by removal of residual propellants, compressed fluids, and electrical storage devices?	Always = 2 Sometimes = 1 Never = 0	1
6. Do the country's laws or administrative regulations or procedures limit the long-term presence of space objects in low-Earth orbit after their end of mission, including through controlled removal from orbit through re-entry or in orbits beyond low-Earth orbit?	Always = 2 Sometimes = 1 Never = 0	1
7. Do the country's laws or administrative regulations or procedures limit the long-term presence of space objects in geosynchronous Earth orbit after their end of mission, including through controlled removal to orbits beyond geosynchronous Earth orbit?	Always = 2 Sometimes = 1 Never = 0	1
Guidelines Score (Sum the above scores and bring to Table B.2)		7

B.2.10	Safety Framework for Nuclear Power Source Applications in Outer Space	Score
1. Does the country have in place safety policies, requirements, and processes that protect people and the environment in Earth's biosphere from potential hazards associated with relevant launch, operation, and end-of-service phases of space nuclear power source applications?	All phases / not applicable = 2 Some phases = 1 No phases = 0	2
2. Does the country's space activities authorisation regime verify the rationale and require justification for the use of space nuclear power source applications?	Always / not applicable = 2 Sometimes = 1 Never = 0	2
3. Does the country's space launch authorisation regime require an independent safety evaluation assessing the risk to people and the environment for launches, operations, and end-of-service phases for space nuclear power source applications?	Always / not applicable = 2 Sometimes = 1 Never = 0	2
4. Does the country conduct emergency preparedness activities, including emergency planning, training, rehearsals and development of procedures and communication protocols, in preparation for radiation exposure to people and the environment as a result of space nuclear power applications?	Repeatedly = 2 Once = 1 Never = 0	2
5. Does the country's laws or administrative regulations or procedures identify space operators of space nuclear power source applications as the primary responsibility holders for operations of such applications?	Full recognition = 2 Partial recognition = 1 No recognition = 0	2
6. Does the country's laws or administrative regulations or procedures require space operators to include safety management to form part of overall space mission management?	Always / not applicable = 2 Sometimes = 1 Never = 0	2
7. Does the country's laws or administrative regulations or procedures require space operations to have technical competence in nuclear safety, including qualified individuals and facilities for designing, testing, and analysing nuclear safety capabilities part of space missions?	Always / not applicable = 2 Sometimes = 1 Never = 0	2
8. Does the country's laws or administrative regulations or procedures require space operators to integrate safety considerations from design to development to launch and operations and end-of-service for the entire space nuclear power source application?	Always / not applicable = 2 Sometimes = 1 Never = 0	2
9. Does the country's laws or administrative regulations or procedures require space operators to conduct risk assessments on the launch, operation, and end-of-service phases of space nuclear power source applications?	Always / not applicable = 2 Sometimes = 1 Never = 0	2
10. Does the country's laws or administrative regulations or procedures require mitigation measures for accidents of space nuclear power source applications with the potential to release radioactive material into the environment?	Always / not applicable = 2 Sometimes = 1 Never = 0	2
Framework Score (Sum the above scores and bring to Table B.2)		20

B.2.11	Voluntary guidelines for transparency and confidence-building, space code of conduct, and long-term sustainability of space	Score
Individual recommendations included within these three separate sets of guidelines are not scored. However, if a country has implemented a majority of the individual recommendations contained within each set of guidelines in either its legal or administrative practice, it will be assessed a score of 2 per set of guidelines. Therefore, the following sets of guidelines for consideration are:		

Appendix II: Scoring for South Korea

i. Recommendations from the Group of Governmental Experts on Transparency and Confidence-Building Measures	0
ii. Recommendations from the European Union's International Code of Conduct on Outer Space Activities	0
iii. Recommendations from the UNCOPUOS Long-Term Sustainability Guidelines	0
Guidelines Score (Sum the above scores and bring to Table B.2)	0

Appendix III: Scoring for South Africa

A	Aggregate Score
Binding International Law Score *(Enter total from last line of Table B.1)*	Emerging: 12.98 / 16.90
Non-Binding International Law Score *(Enter total from last line of Table B.2)*	Emerging: 6.40 / 13.80
Total *(Sum binding and non-binding international law scores)*	Emerging: 19.38 / 30.70

Appendix III: Scoring for South Africa

B.1	Binding International Law Score
Space-specific International Treaties *(Use total from last line of Table B.1.1 and multiply by .50)*	Emerging: 9.88 / 11.0
Space-related International Treaties *(Use total from last line of Table B.1.2 and multiply by .50)*	Emerging: 3.10 / 5.90
Total *(Sum space-specific and space-related treaties scores and then bring total to Table A)*	Emerging: 12.98 / 16.90

B.1.1	Space-specific International Treaties	Score
1. Treaty on Principles Governing the Activities of States in the Exploration and Use of Outer Space, including the Moon and Other Celestial Bodies (25 percent of section score)		
a. Has the country ratified or acceded to this treaty?	Yes, fully = 2 Yes, with reservations = 1 No = 0	2
b. Is the treaty domesticated in national laws or administrative regulations or procedures?	Full domestication = 2 Partial domestication = 1 No domestication = 0	2
c. Article I: Do country's laws or administrative regulations or procedures recognise the role of international law?	Full recognition = 2 Partial recognition = 1 No recognition = 0	2
d. Article I: Do the country's laws or administrative regulations or procedures affirm that the exploration and use of outer space should be carried out for the benefit and interests of all countries?	Full affirmation = 2 Partial affirmation = 1 No affirmation = 0	2
e. Article I: Do the country's laws or administrative regulations or procedures recognise outer space as the province of all mankind?	Full recognition = 2 Partial recognition = 1 No affirmation = 0	0
f. Article I: Do the country's laws or administrative regulations or procedures recognise that outer space should be free for exploration without discrimination and with free access to celestial bodies?	Full recognition = 2 Partial recognition = 1 No recognition = 0	2
g. Article I: Do the country's laws or administrative regulations or procedures recognise the freedom of scientific investigation in outer space?	Full recognition = 2 Partial recognition = 1 No recognition = 0	2
h. Article I: Does the country facilitate and participate in international scientific investigation?	Frequently = 2 Occasionally = 1 Never = 0	2
i. Article II: Do the country's laws or administrative regulations or procedures allow for, or encourage, the appropriation of outer space and celestial bodies by claims of sovereignty, use, or occupation?	Always = 0 Sometimes = 1 Never = 2	
j. Article II: Does the country, by its actions or the actions of those within its jurisdiction, appropriate, or encourage the appropriation of, outer space and celestial bodies and resources by claims of sovereignty, use, or occupation?	Frequently = 0 Occasionally = 1 No = 2	
Article III: Elements of this article are already captured in questions asked above, therefore no separate scoring is assessed.	N/A	
k. Article IV: Has the country placed into orbit or installed on celestial bodies weapons of mass destruction, including nuclear weapons?	Repeatedly = 0 Once = 1 No = 2	
l. Article IV: Has the country established bases, installations, or fortifications, or tested weapons or conducted military manoeuvres on celestial bodies?	Repeatedly = 0 Once = 1 No = 2	
m. Article IV: Do the country's laws or administrative regulations or procedures allow for the conduct of such types of activities as described in Question 'l' above?	Always = 0 Sometimes = 1 Never = 2	
n. Article V: Has the country either provided assistance or rejected assistance to astronauts of other countries who are in distress?	Always provided / not applicable = 2 Sometimes provided = 1 Never provided = 0	2
o. Article V: Does the country proactively inform the United Nations and other countries regarding potentially dangerous phenomena to astronauts?	Always = 2 Sometimes = 1 Never = 0	
p. Article VI: Does the country have in place an authorisation and supervision regime for space activities conducted by non-governmental entities?	Yes = 2 Partial = 1 No = 0	2
q. Article VII: Does the country recognise its liability as a launching state in its laws or administrative regulations or procedures?	Full recognition = 2 Partial recognition = 1 No recognition = 0	
r. Article VII: Has the country denied its liability as a launching state for damage caused?	Always = 0 Sometimes = 1 Never = 2	
s. Article VIII: Does the country maintain an updated national registry of launched space objects?	Always = 2 Sometimes = 1 Never = 0	1
t. Article VIII: Has the country ever returned a space object belonging to another country, but which landed in its jurisdiction?	Always / not applicable = 2 Sometimes = 1 Never = 0	2
u. Article IX: Do the country's laws or administrative regulations or procedures account for avoiding harmful contamination, including debris, of the space environment?	Full account = 2 Partial account = 1 No account = 0	
v. Article IX: Do the country's laws or administrative regulations or procedures seek to limit damage to the Earth environment by extraterrestrial material?	Full account = 2 Partial account = 1 No account = 0	
w. Article X: Does the country allow other countries to observe the flight of space objects that it launches?	Always = 2 Sometimes = 1 Never = 0	
x. Article XI: Does the country inform the United Nations and other countries of its space activities, including the nature, conduct, location, and results of the activities?	Always = 2 Sometimes = 1 Never = 0	1
y. Article XII: Does the country allow access and visits to its outer space stations, installations, equipment, and vehicles, provided reasonable notification of a such visit is given by the visiting representatives?	Always / not applicable = 2 Sometimes = 1 Never = 0	
Articles XIII, XIV, XV, XVI, and XVII: These are administrative articles, therefore no scoring is assessed.		
Treaty Score (Sum the above scores and multiply by .25)		5.5
2. Agreement on the Rescue of Astronauts, the Return of Astronauts and the Return of Objects Launched into Outer Space *(25 percent of section score)*		
	Yes, in full = 2	

Appendix III: Scoring for South Africa

	Scoring	Score
a. Has the country ratified or acceded to this treaty?	Yes, with reservations = 1 No = 0	2
b. Is the treaty domesticated in national laws or administrative regulations or procedures?	Full domestication = 2 Partial domestication = 1 No domestication = 0	0
c. Article 1: Has the country notified the launching state(s) of astronauts who have suffered accidents, been in distress, or made emergency landings?	Always / not applicable = 2 Sometimes = 1 Never = 0	2
d. Article 1: Has the country notified the United Nations of astronauts who have suffered accidents, been in distress, or made emergency landings?	Always / not applicable = 2 Sometimes = 1 No = 0	2
e. Article 2: Has the country taken all possible steps to rescue and render all necessary assistance to astronauts who have suffered accidents, been in distress, or made emergency landings?	Always / not applicable = 2 Sometimes = 1 No = 0	2
f. Article 3: Has the country provided assistance in search and rescue operations for astronauts who have suffered accidents, been in distress, or made emergency landings?	Always / not applicable = 2 Sometimes = 1 No = 0	2
g. Article 4: Has the country safely and promptly returned to their launching states astronauts who have suffered accidents, been in distress, or made emergency landings?	Always / not applicable = 2 Sometimes = 1 No = 0	2
h. Article 5: Has the country notified the launching state and United Nations of a space object or part of a space object that has returned to Earth?	Always / not applicable = 2 Sometimes = 1 No = 0	2
i. Article 5: Has the country in which the returning space object landed practicably assisted to recover the object or its parts?	Always / not applicable = 2 Sometimes = 1 No = 0	2
j. Article 5: Has the country in which the returning space object landed returned the space object or its parts and provided their identifying data to the launching state?	Always / not applicable = 2 Sometimes = 1 No = 0	2
k. Article 5: Has the country in which the returning space object landed eliminated the possibility of its danger if it is found to be hazardous or deleterious?	Always / not applicable = 2 Sometimes = 1 No = 0	2
l. Article 5: Has the launching state paid for expenses related to the recovery and return of space object or its part incurred by the country in which the space object landed?	Always / not applicable = 2 Sometimes = 1 No = 0	2
Articles 6, 7, 8, 9, and 10: These are administrative articles, therefore no scoring is assessed.		
Treaty Score (Sum the above scores and multiply by .25)		5
3. Convention on International Liability for Damage Caused by Space Objects		
a. Has the country ratified or acceded to this treaty?	Yes, in full = 2 Yes, with reservations = 1 No = 0	2
b. Is the treaty domesticated in national laws or administrative regulations or procedures?	Full domestication = 2 Partial domestication = 1 No domestication = 0	2
Article I provides for definitions of terms used in the treaty, therefore no scoring is assessed.		
c. Article II: Do the country's laws or administrative regulations or procedures recognise its liability to pay compensation for damage caused by its space objects on the Earth's surface or aircraft in flight?	Full recognition = 2 Partial recognition = 1 No recognition = 0	2
d. Article II: Has the country refused to pay compensation in practice for damage caused by its space objects on the Earth's surface or aircraft in flight?	Always = 0 Sometimes = 1 Never / not applicable = 2	2
e. Article III: Do the country's laws or administrative regulations or procedures recognise its liability owing to its fault or fault of persons for which it is responsible in the case of damage to another space object, persons, or property in space of another launching state?	Full recognition = 2 Partial recognition = 1 No recognition = 0	2
f. Article III: Has the country refused to accept liability in the case of damage to another space object, persons, or property in space of another launching state caused by its fault or fault of persons for which it is responsible?	Always = 0 Sometimes = 1 Never / not applicable = 2	2
g. Article IV: Do the country's laws or administrative regulations or procedures recognise its absolute joint and several liability with a second-party launching state if its space object caused damage to the second-party launching state's space object, which then resulted in damage on the Earth's surface or aircraft in flight of a third state?	Full recognition = 2 Partial recognition = 1 No recognition = 0	2
h. Article IV: Does the country's laws or administrative regulations or procedures recognise its joint and several liability apportioned according to fault with a second-party launching state if its space object caused damage to the second-party launching state's space object, which then resulted in damage to a third state's space object, person, or property in space?	Full recognition = 2 Partial recognition = 1 No recognition = 0	2
i. Article IV: Has the country refused to pay compensation according to its fault in cases of joint and several liability for damage caused to a third-party state?	Always = 0 Sometimes = 1 Never / not applicable = 2	2
j. Article V: Do the country's laws or administrative regulations or procedures recognise joint and several liability in cases when it and another state jointly launch a space object?	Full recognition = 2 Partial recognition = 1 No recognition = 0	2
k. Article V: Do country's laws or administrative regulations or procedures account for its roles a joint launch participant if a space object is launched from its territory or facility?	Full account = 2 Partial account = 1 No account = 0	2
Articles VI and VII sets forth conditions under which other articles apply and is not actionable by states, therefore no scoring is assessed.		
l. Article VIII: Do the country's laws or administrative regulations or procedures provide for a claims and compensation process for damage which it may suffer by other launching states?	Full provision = 2 Partial provision = 1 No provision = 0	2
Article IX, X, XI, XII, and XIII prescribe the administrative details through which claims and compensation should be processed, therefore no scoring is assessed.		
Articles XIV, XV, XVI, XVII, and XVIII prescribe the administrative details through which a Claims Commission should be established and make adjudications in the case of no settlements through diplomatic channels established elsewhere in the treaty, therefore no scoring is assessed.		
m. Article XIX: Has the country refused or failed to pay compensation or an award decided by a Claims Commission as set forth in other treaty articles?	Always = 0 Sometimes = 1 Never / not applicable = 2	2
Article XX prescribes the process through which the expenses associated with the Claims Commission should be borne by state participants, therefore no scoring is assessed.		
n. Article XXI: Has the country rendered appropriate and rapid assistance in cases when a space object presents large-scale danger to human life or interferes with a population's living conditions or functioning of a vital centre?	Always / not applicable = 2 Sometimes = 1 Never = 0	2
Articles XXII, XXIII, XXIV, XXV, XXVI, XXVII, XXVIII: These are administrative articles, therefore no scoring is assessed.		
Treaty Score (Sum the above scores and multiply by .25)		7
4. Convention on Registration of Objects Launched into Outer Space		
a. Has the country ratified or acceded to this treaty?	Yes, in full = 2 Yes, with reservations = 1 No = 0	2

Appendix III: Scoring for South Africa

	Scoring criteria	Score
b. Is the treaty domesticated in national laws or administrative regulations or procedures?	Full domestication = 2 Partial domestication = 1 No domestication = 0	2
Article I provides for definitions of terms used in the treaty, therefore no scoring is assessed.		
c. Article II: Does the country maintain an updated national registry of launched space objects that it launches and that it launches jointly with other launching states?	Always / not applicable = 2 Sometimes = 1 Never = 0	1
Article III describes actions to be taken by the United Nations rather than a specific country, therefore no scoring is assessed.		
d. Article IV: Does the country provide detailed information to the United Nations for its launched space objects, including name of the country, space object registration number, date and territory of launch, basic orbital parameters, and general function of the space object?	Full information / not applicable= 2 Partial information = 1 No information = 0	1
e. Article IV: Does the country update the United Nations on space objects for which it previously shared information, but which are no longer in Earth orbit?	Always / not applicable = 2 Sometimes = 1 Never = 0	1
Article V describes a procedural administrative action, therefore no scoring is assessed.		
f. Article VI: Does the country assist with providing identifying information it has gained through space monitoring and tracking facilities for space objects that are unidentifiable and which are of a hazardous or deleterious nature?	Always = 2 Sometimes = 1 Never = 0	2
Articles VII, VIII, IX, X, and XII: These are administrative articles, therefore no scoring is assessed.		
Treaty Score (Sum the above scores and multiply by .25)		2.25
Total (Sum the above treaty score and bring total to Table B.1)		19.75

B.1.2	Space-related International Treaties	Score
a. Treaty Banning Nuclear Weapon Tests in the Atmosphere, in Outer Space and under Water		
i. Has the country ratified or acceded to this treaty?	Yes, in full = 2 Yes, with reservations = 1 No = 0	2
ii. Is the treaty domesticated in national laws or administrative regulations or procedures?	Full domestication = 2 Partial domestication = 1 No domestication = 0	2
iii. Article I: Do the country's laws or administrative regulations or procedures prohibit, prevent, and forbid the carrying out of nuclear weapon tests at places under its jurisdiction or control, including in the atmosphere and outer space and underwater?	Full prohibition = 2 Partial prohibition = 1 No prohibition = 0	2
iv. Article I: Since implementing the treaty, has the country carried out a nuclear weapon test at a place under its jurisdiction or control, including in the atmosphere and outer space and underwater?	Repeatedly = 0 Once = 1 Never = 2	2
Articles II, III, IV, and V: These are administrative articles, therefore no scoring is assessed.		
Treaty Score (Sum the above scores)		8
b. Convention Relating to the Distribution of Programme-Carrying Signals Transmitted by Satellite		
i. Has the country ratified or acceded to this treaty?	Yes, in full = 2 Yes, with reservations = 1 No = 0	0
ii. Is the treaty domesticated in national laws or administrative regulations or procedures?	Full domestication = 2 Partial domestication = 1 No domestication = 0	0
Article 1 provides for definitions of terms used in the treaty, therefore no scoring is assessed.		
iii. Article II: Do the country's laws or administrative regulations or procedures prevent the distribution on or from its territory programme-carrying signals by distributors for whom the emitted signal to or passing through the satellite is not intended?	Always = 2 Sometimes = 1 Never = 0	0
iv. Article II: Has the country, by its actions or the actions of those within its jurisdiction, allowed for the distribution on or from its territory programme-carrying signals by distributors for whom the emitted signal to or passing through the satellite is not intended?	Always = 0 Sometimes = 1 Never = 2	0
Articles 3, 4, 5, 6, and 7 describe the conditions in which Article 2 applies, therefore no scoring is assessed.		
Articles 8, 9, 10, 11 and 12: These are administrative articles, therefore no scoring is assessed.		
Treaty Score (Sum the above scores)		0
c. Institution-related treaties		
Individual treaties within this subsection are not scored as they are institution-specific and in some cases outdated and therefore irrelevant. However, collectively if a country ratified or acceded to at least two of these treaties, it will be assessed a score of 2 to demonstrate its participation and cooperation in international legal instruments and organisations related to space, a key principle promoted in international space law documents. Therefore, the following treaties for consideration are:		
i. Agreement Relating to the International Telecommunications Satellite Organization (ITSO)		
ii. Agreement on the Establishment of the INTERSPUTNIK International System and Organization of Space Communications		
iii. Convention for the Establishment of a European Space Agency (ESA)		0.6
iv. Agreement of the Arab Corporation for Space Communications (ARABSAT)		
v. Agreement on Cooperation in the Exploration and Use of Outer Space for Peaceful Purposes (INTERCOSMOS)		
vi. Convention on the International Mobile Satellite Organization		
vii. Convention Establishing the European Telecommunications Satellite Organization (EUTELSAT)		
viii. Convention for the Establishment of a European Organization for the Exploitation of Meteorological Satellites (EUMETSAT)		
ix. International Telecommunication Constitution and Convention		
Total (Sum the treaty scores from (a) and (b) and multiply by .70 and then sum with © score; bring total to Table B.1)		6.2

B.2	Non-Binding International Law Score
(a) Principles Governing the Use by States of Artificial Earth Satellites for International Direct Television Broadcasting *(Enter total from Table B.2.1 and multiply by .10)*	Emerging: 1.60 / 1.80
(b) Principles Relating to Remote Sensing of the Earth from Outer Space *(Enter total from Table B.2.2 and multiply by .10)*	Emerging: 1.70 / 1.80
(c) Principles Relevant to the Use of Nuclear Power Sources in Outer Space *(Enter total from Table B.2.3 and multiply by .10)*	Emerging: 0.0 / 0.0
(d) Declaration on International Cooperation in the Exploration and Use of Outer Space for the Benefit and in the Interest of All States, Taking into Particular Account the Needs of Developing Countries *(Enter total from Table B.2.4 and multiply by .10)*	Emerging: 0.0 / 0.0
(e) International cooperation in the peaceful uses of outer space: Some aspects concerning the use of the geostationary orbit *(Enter total from Table B.2.5 and multiply by .10)*	Emerging: 0.4 /0.4
(f) Application of the concept of the "launching State" *(Enter total from Table B.2.6 and multiply by .10)*	Emerging: 0.6 / 0.6
(g) Recommendations on enhancing the practice of States and international intergovernmental organizations in registering space objects *(Enter total from Table B.2.7 and multiply by .10)*	Emerging: 0.3 / 0.4
(h) Recommendations on national legislation relevant to the peaceful exploration and use of outer space *(Enter total from Table B.2.8 and multiply by .10)*	Emerging: 1.10 / 1.20
(i) Space Debris Mitigation Guidelines of the Committee on the Peaceful Uses of Outer Space *(Enter total from Table B.2.9 and multiply by .10)*	Emerging: 0.7 / 1.40
(j) Safety Framework for Nuclear Power Source Applications in Outer Space *(Enter total from Table B.2.10 and multiply by .10)*	Emerging:0.0 / 0.0
(k) Voluntary guidelines for transparency and confidence-building, space code of conduct, and long-term sustainability of space *(Enter total from Table B.2.11)*	Emerging: 0 / 6
Total *(Sum scores from above and bring total to Table A)*	Emerging: 6.4 / 13.8

B.2.1	Principles Governing the Use by States of Artificial Earth Satellites for International Direct Television Broadcasting	Score
1. Are the country's satellite direct television broadcasting activities respectful of state sovereignty, and do they follow the principle of non-intervention?	Always / not applicable = 2 Sometimes = 1 Never = 0	2
2. Do the country's satellite direct television broadcasting activities promote the free dissemination and exchange of information and knowledge in cultural and scientific fields?	Always / not applicable = 2 Sometimes = 1 Never = 0	2
3. Does the country's satellite direct television broadcasting activities assist in education and social and economic development?	Always / not applicable = 2 Sometimes = 1 Never = 0	2
4. Are the country's satellite direct television broadcasting activities carried out on a friendly and cooperative basis with other countries?	Always / not applicable = 2 Sometimes = 1 Never = 0	0
5. Does the country have in a place an authorisation system through its laws or administrative regulations or procedures for the conduct of satellite direct television activities?	Full implementation = 2 Partial implementation = 1 No implementation = 0	2
6. Does the country consult with other states regarding broadcasting or receiving of satellite direct television broadcasting for purposes of coordinating the same service?	Always / not applicable = 2 Sometimes = 1 Never = 0	2
7. Does the country participate in agreements with other states for the protection of copyright and similar rights?	Always / not applicable = 2 Sometimes = 1 Never = 0	2
8. Does the country notify the United Nations of satellite direct television broadcasting?	Always / not applicable = 2 Sometimes = 1 Never = 0	2
9. Does the country notify and consult the receiving state of a satellite direct television signal of its intent to broadcast?	Always / not applicable = 2 Sometimes = 1 Never = 0	2
Principles Score (Sum the above scores and bring to Table B.2)		16

B.2.2	Principles Relating to Remote Sensing of the Earth from Outer Space	Score
1. Does the country carry out remote sensing activities for the benefit and in the interests of all countries?	Always= 2 Sometimes = 1 Never = 0	1

Appendix III: Scoring for South Africa

	Always = 2 Sometimes = 1 Never = 0	
2. Does the country's remote sensing activities observe the principle of full and permanent sovereignty of all states over their own wealth and natural resources?	Always = 2 Sometimes = 1 Never = 0	1
3. Does the country allow for international cooperation and participation in its remote sensing activities?	Always = 2 Sometimes = 1 Never = 0	2
4. Does the country enter into agreements with other states for the establishment and operation of data collection and storage stations and processing and interpretation facilities?	Always = 2 Sometimes = 1 Never = 0	2
5. Does the country offer technical assistance on a regular basis to other states interested in conducting remote sensing activities?	Always = 2 Sometimes = 1 No = 0	2
6. Does the country inform the United Nations of its remote sensing activities?	Always = 2 Sometimes = 1 Never = 0	1
7. Does the country disclose information to other states derived from remote sensing that could help avert harm to the Earth's natural environment?	Always = 2 Sometimes = 1 Never = 0	2
8. Does the country disclose information to other states derived from processing and analysing remote sensing data that is useful to states affected, or likely to be affected, by natural disasters?	Always = 2 Sometimes = 1 Never = 0	2
9. Does the country make available to remotely sensed states primary and processed data on the sensed state according to non-discriminatory and reasonable cost terms?	Always = 2 Sometimes = 1 Never = 0	2
10. Does the country allow for opportunities for a remotely sensed state to participate and benefit from activities that remotely sense it, including but not limited to access to remotely sensed data and participation in remote sensing activities ?	Always = 2 Sometimes = 1 Never = 0	2
Principles Score (Sum the above scores and bring to Table B.2)		17

B.2.3	**Principles Relevant to the Use of Nuclear Power Sources in Outer Space**	**Score**
1. Does the country design and use space objects with nuclear power sources with high degrees of confidence to avoid hazardous levels associated with operational or accidental circumstances?	Always / not applicable = 2 Sometimes = 1 Never = 0	
2. Do the country's space objects with nuclear power sources follow design and construction that take into account generally accepted international radiological protection guidelines?	Always / not applicable = 2 Sometimes = 1 Never = 0	
3. Do the country's space objects with nuclear power sources possess the capability of correcting or counteracting failures or malfunctions?	Always / not applicable = 2 Sometimes = 1 Never = 0	
4. Do the country's space objects with nuclear power sources include systems that follow practices of redundancy, physical separation, functional isolation, and adequate independence of components?	Always / not applicable = 2 Sometimes = 1 Never = 0	
5. Does the country only operate nuclear power sources on space objects that are on interplanetary missions, in sufficiently high orbits, and in low-Earth orbits that are later stored in sufficiently high orbits after their mission operations?	Always / not applicable= 2 Sometimes = 1 Never = 0	
6. Does the country's space objects with nuclear power sources only use highly enriched uranium 235 as fuel?	Always / not applicable = 2 Sometimes = 1 Never = 0	
7. Does the country only make critical its nuclear reactors once they have reached their operating orbit or interplanetary trajectory?	Always / not applicable= 2 Sometimes = 1 Never = 0	
8. Does the country design and constructs its satellites with nuclear power sources to ensure that they do not become critical before reaching their operating orbit or interplanetary trajectory?	Always / not applicable = 2 Sometimes = 1 Never = 0	
9. Does the country's space objects with nuclear reactors include an effective and controlled disposal of the nuclear reactor for those in an operational orbit less than sufficiently high?	Always / not applicable = 2 Sometimes = 1 Never = 0	
10. Does the country only use radioisotope generators in space objects that are in interplanetary missions, or in missions in Earth orbit but which are later disposed of in high orbit?	Always / not applicable = 2 Sometimes = 1 Never = 0	
11. Does the country's space object with radioisotope generators include containment systems that can withstand the heat and aerodynamic forces of re-entry, and which prevent the scattering of radioactive material in the environment?	Always / not applicable = 2 Sometimes = 1 Never = 0	
12. Does the country conduct a comprehensive safety assessment for all phases of a mission for a space object with a nuclear power source?	Always / not applicable = 2 Sometimes = 1 Never = 0	
13. Does the country indicate to the public and inform the United Nations of when it will launch a space object with a nuclear power source?	Always / not applicable = 2 Sometimes = 1 Never = 0	
14. Does the country provide timely and updated information on space objects with nuclear power sources that malfunction and which may re-enter?	Always / not applicable = 2 Sometimes = 1 Never = 0	
15. Does the country provide assistance to eliminate actual and possible harmful effects of re-entered space objects with nuclear power sources?	Always / not applicable = 2 Sometimes = 1 Never = 0	
Principles Score (Sum the above scores and bring to Table B.2)		0

Appendix III: Scoring for South Africa

B.2.4	Declaration on International Cooperation in the Exploration and Use of Outer Space for the Benefit and in the Interest of All States, Taking into Particular Account the Needs of Developing Countries	Score
1. Does the country engage in equitable and mutually acceptable cooperation with other states according to fair and reasonable contractual terms?	Always = 2 Sometimes = 1 Never = 0	
2. Does the country offer developing countries and countries with incipient space programmes opportunities to participate in advanced space activities?	Always = 2 Sometimes = 1 Never = 0	
3. Do the country's space activities promote the development of space science and technology and its applications?	Always = 2 Sometimes = 1 Never = 0	
4. Do the country's space activities help develop relevant and appropriate capabilities in states interested in space?	Always = 2 Sometimes = 1 Never = 0	
5. Do the country's space activities facilitate the exchange of expertise and technology on a mutually acceptable basis?	Always = 2 Sometimes = 1 No = 0	
Declaration Score (Sum the above scores and bring to Table B.2)		0

B.2.5	International cooperation in the peaceful uses of outer space: Some aspects concerning the use of the geostationary orbit	Score
1. Does the country already having access to a geostationary orbit/spectrum take all practicable steps to enable a developing country, which wishes to use the same, to have equitable access to the requested orbit/spectrum?	Always / not applicable = 2 Sometimes = 1 Never = 0	2
2. Does the country file satellite orbit and frequency requests according to International Telecommunications Union regulations?	Always / not applicable = 2 Sometimes = 1 Never = 0	2
Resolution Score (Sum the above scores and bring to Table B.2)		4

B.2.6	Application of the concept of the "launching State"	Score
1. Does the country have in place national laws authorizing and providing for continuing supervision of non-governmental entities operating in outer space?	Full implementation = 2 Partial implementation = 1 No implementation = 0	2
2. Does the country voluntarily provide information for the on-orbit transfer of ownership of space objects?	Always = 2 Sometimes = 1 Never = 0	2
3. Does the country have harmonized national space laws with international space laws?	Full harmonization = 2 Partial harmonization = 1 No harmonization = 0	2
Resolution Score (Sum the above scores and bring to Table B.2)		6

B.2.7	Recommendations on enhancing the practice of States and international intergovernmental organizations in registering space objects	Score
1. Does the country provide detailed information on its registered space objects to the United Nations, including: international designators, Coordinated Universal Time of launch date, kilometres and minutes and degrees of orbital parameters, geostationary orbit location (as applicable), change in status of operations, date decay or re-entry, date and condition of moving to a disposal orbit, and web links to space object information?	All information = 1 Partial information = 0 No information = -1	
2. Does the country from which the territory or facility a space object is launched jointly determine with other states or intergovernmental organisations which entity should register the space object?	Always / not applicable = 2 Sometimes = 1 Never = 0	
3. If participating in a joint launch, does the country register the space object separately from its joint launching state?	Always / not applicable = 2 Sometimes = 1 Never = 0	1
4. If a space object changes supervision, does the country of registry provide to the United Nations detailed information, including date of change in supervision, identification of the new owner or operator, change in orbital position, and change of function?	All information = 2 Partial information = 1 No information = 0	2
Resolution Score (Sum the above scores and bring to Table B.2)		3

B.2.8	Recommendations on national legislation relevant to the peaceful exploration and use of outer space	Score
1. Does the country have a national framework that governs the launch of objects into and their return from outer space, the operation of launch or re-entry sites and the operation and control of space objects in orbit, design and manufacture of spacecraft, application of space science and technology, and exploration activities and research?	All = 2 Some = 1 None = 0	
2. Does the country issue authorisations and ensure supervision over space activities carried out from its jurisdiction or elsewhere when carried out by its citizens or legally established persons?	Always = 2 Sometimes = 1 Never = 0	2
3. Does the country have a national authority that authorizes space activities, including procedures for granting, modifying, suspending, and revoking authorisations?	All procedures = 2 Some procedures = 1 No procedures = 0	2

			Score
4. Does the country's authorisation regime ensure the safe conduct and minimal risk to persons for space activities, including assessing the expertise and technical qualifications of space operators and requiring adherence to safety and technical standards in line with the Space Debris Mitigation Guidelines?	Fully ensures = 2 Partially ensures = 1 Does not ensure = 0		1
5. Does the country's supervision and authorisation regime include on-site inspections or general reporting requirements?	Fully includes = 2 Partially includes = 1 Does not include = 0		2
6. Does the country's laws or administrative regulations or procedures offer recourse from operators or owners of space objects if the country's liability is engaged in the case of damages, such as through the use of insurance requirements or indemnification procedures?	Full recourse = 2 Partial recourse = 1 No recourse = 0		2
7. Does the country's supervision regime apply in case of transfer of ownership of space objects of non-governmental entities?	Always / not applicable = 2 Sometimes = 1 Never = 0		2
Recommendations Score (Sum the above scores and bring to Table B.2)			11

B.2.9	**Space Debris Mitigation Guidelines of the Committee on the Peaceful Uses of Outer Space**	Score
1. Do the country's laws or administrative regulations or procedures require space systems to be designed to not release debris during normal operations?	Always = 2 Sometimes = 1 Never = 0	1
2. Do the country's laws or administrative regulations or procedures require spacecraft and launch vehicle orbital stages to be designed to avoid accidental break-ups?	Always = 2 Sometimes = 1 Never = 0	1
3. Do the country's laws or administrative regulations or procedures require spacecraft and launch vehicle orbital stages to have disposal and passivation measures in the case of failure?	Always = 2 Sometimes = 1 Never = 0	1
4. Do the country's laws or administrative regulations or procedures require spacecraft operators to follow collision avoidance procedures through adjustment of launch times and on-orbit avoidance manoeuvres?	Always = 2 Sometimes = 1 Never = 0	1
5. Does the country's laws or administrative regulations or procedures require space objects to deplete or make safe their on-board sources of stored energy, including through passivation by removal of residual propellants, compressed fluids, and electrical storage devices?	Always = 2 Sometimes = 1 Never = 0	1
6. Do the country's laws or administrative regulations or procedures limit the long-term presence of space objects in low-Earth orbit after their end of mission, including through controlled removal from orbit through re-entry or in orbits beyond low-Earth orbit?	Always = 2 Sometimes = 1 Never = 0	1
7. Do the country's laws or administrative regulations or procedures limit the long-term presence of space objects in geosynchronous Earth orbit after their end of mission, including through controlled removal to orbits beyond geosynchronous Earth orbit?	Always = 2 Sometimes = 1 Never = 0	1
Guidelines Score (Sum the above scores and bring to Table B.2)		7

B.2.10	**Safety Framework for Nuclear Power Source Applications in Outer Space**	Score
1. Does the country have in place safety policies, requirements, and processes that protect people and the environment in Earth's biosphere from potential hazards associated with relevant launch, operation, and end-of-service phases of space nuclear power source applications?	All phases / not applicable = 2 Some phases = 1 No phases = 0	
2. Does the country's space activities authorisation regime verify the rationale and require justification for the use of space nuclear power source applications?	Always / not applicable = 2 Sometimes = 1 Never = 0	
3. Does the country's space launch authorisation regime require an independent safety evaluation assessing the risk to people and the environment for launches, operations, and end-of-service phases for space nuclear power source applications?	Always / not applicable = 2 Sometimes = 1 Never = 0	
4. Does the country conduct emergency preparedness activities, including emergency planning, training, rehearsals and development of procedures and communication protocols, in preparation for radiation exposure to people and the environment as a result of space nuclear power applications?	Repeatedly = 2 Once = 1 Never = 0	
5. Does the country's laws or administrative regulations or procedures identify space operators of space nuclear power source applications as the primary responsibility holders for operations of such applications?	Full recognition = 2 Partial recognition = 1 No recognition = 0	
6. Does the country's laws or administrative regulations or procedures require space operators to include safety management to form part of overall space mission management?	Always / not applicable = 2 Sometimes = 1 Never = 0	
7. Does the country's laws or administrative regulations or procedures require space operations to have technical competence in nuclear safety, including qualified individuals and facilities for designing, testing, and analysing nuclear safety capabilities part of space missions?	Always / not applicable = 2 Sometimes = 1 Never = 0	
8. Does the country's laws or administrative regulations or procedures require space operators to integrate safety considerations from design to development to launch and operations and end-of-service for the entire space nuclear power source application?	Always / not applicable = 2 Sometimes = 1 Never = 0	
9. Does the country's laws or administrative regulations or procedures require space operators to conduct risk assessments on the launch, operation, and end-of-service phases of space nuclear power source applications?	Always / not applicable = 2 Sometimes = 1 Never = 0	
10. Does the country's laws or administrative regulations or procedures require mitigation measures for accidents of space nuclear power source applications with the potential to release radioactive material into the environment?	Always / not applicable = 2 Sometimes = 1 Never = 0	
Framework Score (Sum the above scores and bring to Table B.2)		0

B.2.11	**Voluntary guidelines for transparency and confidence-building, space code of conduct, and long-term sustainability of space**	Score
Individual recommendations included within these three separate sets of guidelines are not scored. However, if a country has implemented a majority of the individual recommendations contained within each set of guidelines in either its legal or administrative practice, it will be assessed a score of 2 per set of guidelines. Therefore, the following sets of guidelines for consideration are:		

Appendix III: Scoring for South Africa

i. Recommendations from the Group of Governmental Experts on Transparency and Confidence-Building Measures		0
ii. Recommendations from the European Union's International Code of Conduct on Outer Space Activities		0
iii. Recommendations from the UNCOPUOS Long-Term Sustainability Guidelines		0
Guidelines Score (Sum the above scores and bring to Table B.2)		0